ADDING & SUBTRACTING F

Write each sum or difference in its simplest form.

$$+\frac{\frac{7}{12}}{\frac{11}{12}}$$

$$-\frac{\frac{9}{11}}{\frac{3}{11}}$$

$$-\frac{\frac{11}{15}}{\frac{7}{15}}$$

$$+\frac{\frac{13}{15}}{\frac{7}{15}}$$

$$\frac{3}{8} = \frac{3}{8}$$
$$+\frac{1}{4} = +\frac{2}{8}$$
$$\frac{5}{8}$$

$$-\frac{\frac{13}{14}}{\frac{3}{14}}$$

$$-\frac{\frac{6}{7}}{\frac{2}{7}}$$

$$+\frac{\frac{1}{16}}{\frac{3}{16}}$$

$$-\frac{\frac{3}{20}}{\frac{1}{20}}$$

$$\frac{1}{2} = \frac{3}{6}$$
$$-\frac{1}{6} = -\frac{1}{6}$$
$$\frac{2}{6} = \frac{1}{3}$$

$$+\frac{\frac{3}{8}}{\frac{1}{4}}$$

$$-\frac{\frac{1}{2}}{\frac{1}{6}}$$

$$-\frac{\frac{4}{9}}{\frac{1}{3}}$$

$$+\frac{\frac{3}{8}}{\frac{1}{4}}$$

$$+\frac{\frac{2}{3}}{\frac{1}{4}}$$

$$-\frac{\frac{7}{16}}{\frac{3}{8}}$$

$$+\frac{\frac{4}{5}}{\frac{2}{3}}$$

$$-\frac{\frac{3}{4}}{\frac{1}{2}}$$

$$+\frac{\frac{3}{4}}{\frac{2}{5}}$$

$$-\frac{\frac{9}{14}}{\frac{1}{7}}$$

$$+\frac{\frac{5}{8}}{\frac{1}{4}}$$

$$+\frac{\frac{5}{6}}{\frac{1}{12}}$$

1

FRACTIONS

ADDING MIXED NUMBERS.

EXAMPLE:

$$8\frac{7}{10} = \quad 8\frac{7}{10}$$
$$+16\frac{1}{2} = \quad +16\frac{5}{10}$$

Write each sum in its simplest form.

$$24\frac{12}{10} = 25\frac{2}{10} = 25\frac{1}{5}$$

$$8\frac{7}{10}$$
$$+16\frac{1}{2}$$

$$3\frac{1}{4}$$
$$+18\frac{1}{2}$$

$$4\frac{1}{8}$$
$$+16\frac{1}{4}$$

$$3\frac{1}{8}$$
$$+16\frac{1}{2}$$

$$8\frac{1}{5}$$
$$+14\frac{1}{2}$$

$$7\frac{1}{3}$$
$$+17\frac{1}{4}$$

$$18\frac{1}{8}$$
$$+6\frac{9}{16}$$

$$71\frac{1}{16}$$
$$+56\frac{1}{2}$$

$$29\frac{3}{4}$$
$$+15\frac{1}{16}$$

$$52\frac{1}{2}$$
$$+63\frac{3}{4}$$

$$27\frac{3}{4}$$
$$+46\frac{1}{8}$$

$$60\frac{1}{2}$$
$$+69\frac{5}{16}$$

$$50\frac{2}{3}$$
$$+76\frac{1}{6}$$

$$91\frac{1}{2}$$
$$+52\frac{5}{8}$$

$$84\frac{5}{6}$$
$$+94\frac{2}{3}$$

$$78\frac{2}{3}$$
$$+27\frac{3}{4}$$

$$37\frac{2}{3}$$
$$+19\frac{1}{6}$$

ADDING MIXED NUMBERS.

Write each sum in its simplest form.

$$35\frac{7}{8} = 35\frac{7}{8}$$
$$+36\frac{1}{2} = +36\frac{4}{8}$$
$$71\frac{11}{8} = 72\frac{3}{8}$$

$35\frac{7}{8}$
$+36\frac{1}{2}$
_____

$13\frac{1}{6}$
$+\ 8\frac{1}{4}$
_____

$25\frac{9}{10}$
$+\ 5\frac{3}{5}$
_____

$4\frac{1}{3}$
$+24\frac{1}{6}$
_____

$11\frac{1}{6}$
$+\ 9\frac{1}{2}$
_____

$9\frac{1}{2}$
$+13\frac{1}{3}$
_____

$68\frac{1}{2}$
$+25\frac{3}{8}$
_____

$72\frac{3}{4}$
$+67\frac{5}{8}$
_____

$7\frac{1}{3}$
$+19\frac{3}{4}$
_____

$14\frac{2}{3}$
$+\ 7\frac{1}{2}$
_____

$5\frac{4}{9}$
$+21\frac{2}{3}$
_____

$74\frac{1}{4}$
$+43\frac{3}{8}$
_____

$12\frac{7}{12}$
$+26\frac{3}{4}$
_____

$50\frac{9}{10}$
$+17\frac{1}{2}$
_____

$37\frac{5}{8}$
$+49\frac{1}{4}$
_____

$9\frac{1}{3}$
$+17\frac{5}{8}$
_____

$29\frac{5}{12}$
$+39\frac{1}{2}$
_____

$45\frac{1}{2}$
$+27\frac{2}{5}$
_____

FRACTIONS

ADDING MIXED NUMBERS.

Write each sum in its simplest form.

EXAMPLE:

$7\frac{3}{7}$

$+ \ 4\frac{5}{7}$

$11\frac{8}{7} = 12\frac{1}{7}$

$7\frac{3}{7}$
$+ \ 4\frac{5}{7}$

$2\frac{1}{9}$
$+ \ 5\frac{1}{3}$

$3\frac{7}{10}$
$+ \ 8\frac{3}{10}$

$11\frac{6}{7}$
$+ \ 9\frac{2}{7}$

$12\frac{1}{2}$
$+10\frac{3}{7}$

$11\frac{3}{4}$
$+ \ 8\frac{2}{5}$

$6\frac{7}{8}$
$+ \ 1\frac{1}{4}$

$23\frac{2}{9}$
$+ \ 5\frac{7}{9}$

$17\frac{3}{4}$
$+18\frac{1}{5}$

$16\frac{3}{5}$
$+ \ 4\frac{2}{5}$

$15\frac{7}{12}$
$+ \ 7\frac{7}{12}$

$18\frac{10}{11}$
$+ \ 5\frac{9}{11}$

$14\frac{4}{5}$
$+ \ 5\frac{2}{3}$

$8\frac{1}{3}$
$+ \ 4\frac{5}{7}$

$26\frac{9}{10}$
$+14\frac{1}{10}$

4

$\mathcal{S}$OLVING STORY PROBLEMS.

1. Chandra and Michelle each needed $50.00 to go to a basketball camp. Chandra had $\frac{3}{5}$ of the money she needed. Michelle had $\frac{7}{10}$ of the money she needed. How much money did each girl have? Chandra:_____ Michelle: _____

2. During the morning Eric completed $\frac{3}{8}$ of his math problems. In the afternoon he completed another $\frac{1}{4}$ of the problems. How much of the math assignment did Eric have completed when he left school? _____

3. Chandra worked $15\frac{5}{6}$ minutes on the exercise bike on Monday and $12\frac{2}{3}$ minutes on Wednesday. How much total time did Chandra spend on the bike? _____

4. Justine swam for $2\frac{1}{4}$ hours on Saturday and $3\frac{1}{3}$ hours on Sunday. How much longer did she swim on Sunday than on Saturday? _____
How much time did Justine spend swimming this weekend? _____

5. Tom rides his bicycle $2\frac{1}{7}$ miles to school each day. How many miles will Tom ride each day to and from school? _____ In a week, how far does he ride? _____

6. Michelle has a big problem. She needs to simplify the following fractions, but she isn't sure what to do. Simplify these fractions for Michelle:

$\frac{9}{12}$ = $\frac{14}{18}$ = $\frac{12}{20}$ = $\frac{10}{12}$ = $\frac{8}{24}$ =

7. Keith needed $5\frac{1}{2}$ pounds of nails to build a shed. The hardware store had only $3\frac{1}{8}$ pounds of nails. How many more pounds of nails did Keith need? _____

8. Mrs. McJenkins' class was making a stew. Angela brought $3\frac{1}{2}$ lbs. of meat, Kevin brought $1\frac{1}{4}$ lbs. of carrots, Heather brought $\frac{3}{8}$ lbs. of celery and Nathan brought $6\frac{1}{2}$ lbs. of potatotes. How many pounds of ingredients were there? _____

FRACTIONS

ADD & SUBTRACT FRACTIONS.

EXAMPLES:

$$\frac{1}{2} = \frac{3}{6}$$
$$+\frac{1}{3} = +\frac{2}{6}$$
$$\frac{5}{6}$$

$$7\frac{1}{2} = 7\frac{4}{8}$$
$$-4\frac{1}{8} = -4\frac{1}{8}$$
$$3\frac{3}{8}$$

Write each sum or difference in its simplest form.

$$\frac{1}{2}$$
$$-\frac{1}{6}$$

$$\frac{1}{6}$$
$$+\frac{1}{3}$$

$$\frac{5}{6}$$
$$-\frac{1}{2}$$

$$\frac{1}{2}$$
$$+\frac{1}{4}$$

$$7\frac{1}{2}$$
$$-4\frac{1}{8}$$

$$8\frac{1}{2}$$
$$+3\frac{5}{8}$$

$$7\frac{2}{3}$$
$$-3\frac{3}{6}$$

$$5\frac{1}{2}$$
$$+6\frac{3}{4}$$

$$3\frac{1}{3}$$
$$-1\frac{1}{4}$$

$$8\frac{5}{6}$$
$$-3\frac{1}{3}$$

$$13\frac{2}{5}$$
$$-6\frac{1}{10}$$

$$15\frac{1}{2}$$
$$-9\frac{3}{8}$$

$$29\frac{5}{6}$$
$$+3\frac{1}{2}$$

$$5\frac{2}{3}$$
$$-3\frac{1}{6}$$

$$8\frac{1}{10}$$
$$+9\frac{1}{2}$$

6

Subtracting Mixed Numbers.

Write each difference in its simplest form.

$$20$$
$$- \ 5\frac{3}{4}$$

$$50\frac{5}{6}$$
$$- \quad \frac{1}{6}$$

$$3\frac{7}{8}$$
$$- \quad \frac{3}{4}$$

$$5\frac{4}{5}$$
$$- \quad \frac{1}{2}$$

$$57$$
$$- 18\frac{2}{3}$$

$$5\frac{3}{4}$$
$$- \quad \frac{4}{5}$$

$$2\frac{1}{3}$$
$$- \quad \frac{2}{3}$$

$$15$$
$$- \ 6\frac{2}{5}$$

$$9\frac{1}{5}$$
$$- \quad \frac{2}{3}$$

$$4\frac{1}{3}$$
$$- \quad \frac{1}{2}$$

$$99$$
$$-88\frac{3}{8}$$

$$4\frac{1}{2}$$
$$- \quad \frac{5}{6}$$

$$2\frac{5}{12}$$
$$- \quad \frac{3}{4}$$

$$9\frac{1}{3}$$
$$- \quad \frac{5}{6}$$

FRACTIONS

SUBTRACTING MIXED NUMBERS.

EXAMPLE:

$$1\frac{1}{6} = \frac{7}{6}$$
$$-\frac{5}{6} = -\frac{5}{6}$$
$$\frac{2}{6} = \frac{1}{3}$$

Write each difference in its simplest form.

$38\frac{5}{7}$ $-\frac{3}{7}$	$1\frac{1}{6}$ $-\frac{5}{6}$	$3\frac{3}{8}$ $-\frac{5}{8}$	
$4\frac{1}{4}$ $-\frac{3}{4}$	$8\frac{1}{10}$ $-\frac{1}{2}$	$5\frac{2}{5}$ $-\frac{9}{10}$	$7\frac{1}{6}$ $-\frac{2}{3}$
$2\frac{2}{9}$ $-\frac{1}{3}$	$7\frac{3}{4}$ $-\frac{1}{2}$	$8\frac{9}{10}$ $-\frac{2}{5}$	$4\frac{2}{3}$ $-\frac{1}{6}$
$4\frac{1}{5}$ $-\frac{3}{10}$	$12\frac{1}{10}$ $-\frac{4}{5}$	$3\frac{2}{7}$ $-\frac{1}{2}$	$6\frac{1}{6}$ $-\frac{1}{3}$

8

$\int$UBTRACTING MIXED NUMBERS.

Write each difference in its simplest form.

$52\frac{5}{8}$ $-\ 27$	12 $-\ 3\frac{3}{5}$	92 $-\ 66\frac{1}{6}$
$54\frac{5}{6}$ $-\ 18$	33 $-\ 27\frac{5}{8}$	$78\frac{3}{5}$ $-\ 49$	$13\frac{5}{12}$ $-\ 7$
19 $-\ 6\frac{9}{10}$	72 $-\ 53\frac{2}{7}$	93 $-\ 47\frac{4}{9}$	25 $-\ 18\frac{1}{2}$
$61\frac{2}{5}$ $-\ 25$	46 $-\ 11\frac{6}{7}$	$84\frac{1}{4}$ $-\ 56$	$26\frac{7}{8}$ $-\ 18$
$13\frac{3}{7}$ $-\ 2$	$45\frac{2}{3}$ $-\ 36$	$82\frac{4}{7}$ $-\ 8$	40 $-\ 12\frac{1}{5}$

EXAMPLE:

$$24\frac{2}{15} = 23\frac{17}{15}$$
$$- \quad 6\frac{11}{15} = - \quad 6\frac{11}{15}$$
$$\overline{\qquad\qquad 17\frac{6}{15} = 17\frac{2}{5}}$$

Write each difference
in its simplest form.

$$10\frac{1}{5}$$
$$- \quad 6\frac{3}{5}$$

$$5\frac{3}{7}$$
$$- \quad 1\frac{6}{7}$$

$$11\frac{5}{12}$$
$$- \quad 3\frac{11}{12}$$

$$6\frac{3}{7}$$
$$- \quad 4\frac{5}{7}$$

$$9\frac{2}{5}$$
$$- \quad 5\frac{3}{5}$$

$$8\frac{3}{10}$$
$$- \quad 2\frac{9}{10}$$

$$21\frac{2}{15}$$
$$- \quad 8\frac{7}{15}$$

$$12\frac{1}{12}$$
$$- \quad 5\frac{7}{12}$$

$$7\frac{1}{10}$$
$$- \quad 6\frac{3}{10}$$

$$15\frac{3}{8}$$
$$- 11\frac{7}{8}$$

$$30\frac{1}{8}$$
$$- 17\frac{7}{8}$$

$$25\frac{4}{9}$$
$$- \quad 6\frac{7}{9}$$

$$24\frac{3}{14}$$
$$- 23\frac{11}{14}$$

$$18\frac{2}{7}$$
$$- 13\frac{6}{7}$$

Subtract and simplify these fractions.

82 $-\ \frac{6}{8}$	$10\frac{3}{6}$ $-\ 2\frac{4}{6}$	$11\frac{2}{10}$ $-\ 3\frac{4}{10}$	$14\frac{2}{7}$ $-\ 7\frac{6}{7}$
$62\frac{5}{12}$ $-\ 40\frac{7}{12}$	$14\frac{2}{7}$ $-\ 6\frac{6}{7}$	$5\frac{3}{10}$ $-\ 2\frac{7}{10}$	$9\frac{3}{8}$ $-\ 4\frac{7}{8}$
$10\frac{1}{5}$ $-\ 5\frac{3}{5}$	$6\frac{1}{8}$ $-\ 3\frac{4}{8}$	$18\frac{2}{16}$ $-\ 17\frac{5}{16}$	$22\frac{2}{14}$ $-\ 20\frac{3}{14}$
$85\frac{1}{6}$ $-\ 1\frac{9}{6}$	$2\frac{8}{10}$ $-\ 1\frac{9}{10}$	$6\frac{1}{5}$ $-\ 2\frac{4}{5}$	$4\frac{1}{8}$ $-\ 2\frac{3}{8}$
$6\frac{1}{9}$ $-\ 2\frac{6}{9}$	18 $-\ \frac{2}{8}$	$4\frac{5}{10}$ $-\ 2\frac{9}{10}$	6 $-\ 3\frac{5}{7}$
$34\frac{7}{10}$ $-\ 30\frac{9}{10}$	$20\frac{8}{12}$ $-\ 10\frac{9}{12}$	$5\frac{1}{10}$ $-\ 1\frac{3}{10}$	$39\frac{3}{8}$ $-\ 29\frac{5}{8}$

11

FRACTIONS

⌡UBTRACTING MIXED NUMBERS.

Write each
difference
in its simplest form.

$$4\frac{1}{2}$$
$$-\ 2\frac{5}{8}$$

$$19\frac{4}{5}$$
$$-\ 8\frac{3}{10}$$

$$16\frac{5}{6}$$
$$-12\frac{1}{3}$$

$$31\frac{7}{10}$$
$$-10\frac{1}{2}$$

$$7\frac{1}{12}$$
$$-\ 3\frac{1}{6}$$

$$8\frac{1}{3}$$
$$-\ 1\frac{3}{5}$$

$$22\frac{2}{5}$$
$$-10\frac{1}{2}$$

$$10\frac{1}{8}$$
$$-\ 6\frac{1}{2}$$

$$11\frac{1}{9}$$
$$-\ 6\frac{2}{3}$$

$$9\frac{2}{15}$$
$$-\ 4\frac{4}{5}$$

$$12\frac{1}{4}$$
$$-\ 7\frac{3}{8}$$

$$5\frac{1}{6}$$
$$-\ 2\frac{5}{9}$$

$$15\frac{5}{8}$$
$$-11\frac{1}{12}$$

$$12\frac{3}{4}$$
$$-\ 5\frac{2}{3}$$

$$3\frac{1}{2}$$
$$-\ 1\frac{1}{4}$$

$$6\frac{3}{10}$$
$$-\ 2\frac{1}{2}$$

SUBTRACTING MIXED NUMBERS.

EXAMPLE:

$$7\frac{1}{6} = 7\frac{1}{6} = 6\frac{7}{6}$$
$$-5\frac{1}{2} = -5\frac{3}{6} = -5\frac{3}{6}$$
$$1\frac{4}{6} = 1\frac{2}{3}$$

Write each difference in its simplest form.

$7\frac{3}{8}$ $-2\frac{1}{2}$	$5\frac{1}{5}$ $-1\frac{1}{2}$	$4\frac{1}{8}$ $-2\frac{1}{4}$	$6\frac{1}{3}$ $-2\frac{3}{4}$
$6\frac{1}{5}$ $-3\frac{3}{10}$	$5\frac{1}{4}$ $-2\frac{1}{2}$	$4\frac{3}{10}$ $-1\frac{4}{5}$	$8\frac{1}{3}$ $-2\frac{5}{6}$
$7\frac{1}{3}$ $-4\frac{8}{15}$	$8\frac{1}{5}$ $-6\frac{7}{10}$	$9\frac{3}{4}$ $-4\frac{11}{12}$	$7\frac{1}{6}$ $-5\frac{1}{2}$
$6\frac{4}{9}$ $-3\frac{2}{3}$	$3\frac{3}{8}$ $-1\frac{1}{2}$	$9\frac{1}{7}$ $-6\frac{3}{14}$	$4\frac{1}{8}$ $-3\frac{3}{4}$

FRACTIONS

SUBTRACTING MIXED NUMBERS.

Write each difference in its simplest form.

$$10\tfrac{1}{3}$$
$$-\ 8\tfrac{11}{15}$$
_____

$$8\tfrac{1}{6}$$
$$-\ 5\tfrac{1}{3}$$
_____

$$9\tfrac{1}{12}$$
$$-\ 3\tfrac{3}{4}$$
_____

$$4\tfrac{1}{5}$$
$$-\ 2\tfrac{3}{10}$$
_____

$$11\tfrac{2}{5}$$
$$-\ 7\tfrac{7}{10}$$
_____

$$12\tfrac{4}{9}$$
$$-\ 4\tfrac{2}{3}$$
_____

$$4\tfrac{1}{6}$$
$$-\ 3\tfrac{2}{3}$$
_____

$$6\tfrac{1}{12}$$
$$-\ 4\tfrac{5}{6}$$
_____

$$9\tfrac{1}{4}$$
$$-\ 5\tfrac{1}{3}$$
_____

$$8\tfrac{1}{2}$$
$$-\ 3\tfrac{7}{8}$$
_____

$$5\tfrac{1}{2}$$
$$-\ 2\tfrac{5}{6}$$
_____

$$6\tfrac{3}{10}$$
$$-\ 1\tfrac{1}{2}$$
_____

$$16\tfrac{1}{4}$$
$$-\ 9\tfrac{3}{4}$$
_____

$$8\tfrac{3}{10}$$
$$-\ 4\tfrac{7}{10}$$
_____

$$7\tfrac{5}{12}$$
$$-\ 2\tfrac{7}{12}$$
_____

$$5\tfrac{1}{7}$$
$$-\ 2\tfrac{1}{2}$$
_____

$$15\tfrac{1}{6}$$
$$-\ 8\tfrac{5}{6}$$
_____

14

SUBTRACTING WITH FRACTIONS.

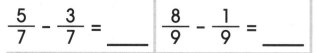

Write each difference in its simplest form.

$$\frac{5}{7} - \frac{3}{7} = \boxed{\frac{2}{7}}$$

$$\begin{array}{ll} 5\frac{1}{9} = & 4\frac{10}{9} \\ -\ 1\frac{7}{9} = & -\ 1\frac{7}{9} \\ \hline & \end{array}$$

$$3\frac{3}{9} = 3\frac{1}{3}$$

$\dfrac{5}{7} - \dfrac{3}{7} = \underline{\hspace{1cm}}$ $\dfrac{8}{9} - \dfrac{1}{9} = \underline{\hspace{1cm}}$

$\dfrac{9}{20} - \dfrac{3}{20} = \underline{\hspace{1cm}}$ $\dfrac{13}{18} - \dfrac{5}{18} = \underline{\hspace{1cm}}$

$\dfrac{8}{15} - \dfrac{1}{15} = \underline{\hspace{1cm}}$ $\dfrac{4}{9} - \dfrac{1}{9} = \underline{\hspace{1cm}}$

$\dfrac{9}{10} - \dfrac{3}{10} = \underline{\hspace{1cm}}$ $\dfrac{5}{21} - \dfrac{2}{21} = \underline{\hspace{1cm}}$

$\dfrac{9}{14} - \dfrac{5}{14} = \underline{\hspace{1cm}}$ $\dfrac{4}{15} - \dfrac{2}{15} = \underline{\hspace{1cm}}$ $\dfrac{11}{16} - \dfrac{5}{16} = \underline{\hspace{1cm}}$ $\dfrac{6}{7} - \dfrac{4}{7} = \underline{\hspace{1cm}}$

$$\begin{array}{l} 6\frac{4}{5} \\ -\ 2\frac{1}{5} \\ \hline \end{array} \qquad \begin{array}{l} 4\frac{1}{8} \\ -\ 2\frac{5}{8} \\ \hline \end{array} \qquad \begin{array}{l} 10\frac{2}{3} \\ -\ 3\frac{1}{3} \\ \hline \end{array} \qquad \begin{array}{l} 8\frac{13}{14} \\ -\ 8\frac{3}{14} \\ \hline \end{array}$$

$$\begin{array}{l} 3\frac{1}{7} \\ -\ 1\frac{5}{7} \\ \hline \end{array} \qquad \begin{array}{l} 7\frac{3}{10} \\ -\ 5\frac{1}{10} \\ \hline \end{array} \qquad \begin{array}{l} 12\frac{7}{15} \\ -\ 4\frac{4}{15} \\ \hline \end{array} \qquad \begin{array}{l} 5\frac{7}{10} \\ -\ 2\frac{9}{10} \\ \hline \end{array}$$

$$\begin{array}{l} 14\frac{3}{7} \\ -11\frac{1}{7} \\ \hline \end{array} \qquad \begin{array}{l} 23\frac{7}{8} \\ -\ 9\frac{5}{8} \\ \hline \end{array} \qquad \begin{array}{l} 15\frac{11}{12} \\ -\ 8\frac{7}{12} \\ \hline \end{array} \qquad \begin{array}{l} 11\frac{4}{9} \\ -\ 3\frac{5}{9} \\ \hline \end{array}$$

FRACTIONS

$\int$OLVING STORY PROBLEMS.

1. Donna ate $\frac{1}{2}$ of the cookies and Paul ate $\frac{3}{8}$ of the cookies. What fraction of the cookies did they eat? _____

2. Tom painted $\frac{1}{4}$ of the fence and Kim painted $\frac{3}{12}$ of the fence. How much of the fence has been painted? _____

3. Barry wrote $\frac{7}{10}$ of the spelling words correctly. Diane wrote $\frac{3}{5}$ of the words correctly. Who spelled more words correctly? _____ How many more? _____

4. Susan skipped $\frac{1}{3}$ of the way home, ran $\frac{1}{6}$ of the way home, and walked $\frac{1}{2}$ of the way home. Did she reach home? _____ Show the fraction. _____

5. Anita cut a pie into 8 pieces. She ate 1 piece, Dan ate 2 pieces, and Dora ate 2 pieces. What fraction of the pie did they eat? _____

6. Jerry, Barry, and Larry each painted $\frac{1}{4}$ of the garage. How much of the garage did not get painted? _____

7. Tyresha lives $\frac{9}{10}$ of a mile from school. Fred lives $\frac{4}{5}$ of a mile from school. Who lives the farthest from school? _____ How much farther? _____

8. Bill put $1\frac{1}{4}$ cups of milk and $\frac{3}{16}$ cup of oil into the pancake batter. How much liquid did he put in? _____

9. Jerome raked $\frac{1}{3}$ of the yard in the morning and $\frac{1}{3}$ of the yard in the afternoon. How much of the yard did not get raked? _____

10. Anna was assigned to read 15 pages. She read $\frac{1}{3}$ of the pages in class and $\frac{2}{5}$ more pages in study period. How many more pages were left to read? _____

16

MULTIPLYING WITH FRACTIONS.

Write each product in its simplest form.

EXAMPLE:

$$\frac{5}{\cancel{6}_{1}} \times \frac{\overset{4}{\cancel{24}}}{1} = 20$$

$\frac{2}{3} \times 36 = \underline{\quad}$ $\frac{5}{8} \times 48 = \underline{\quad}$ $\frac{2}{5} \times 45 = \underline{\quad}$

$\frac{3}{4} \times 48 = \underline{\quad}$ $\frac{5}{6} \times 54 = \underline{\quad}$ $\frac{4}{7} \times 49 = \underline{\quad}$ $\frac{3}{4} \times 72 = \underline{\quad}$

$\frac{2}{3} \times 18 = \underline{\quad}$ $\frac{3}{4} \times 24 = \underline{\quad}$ $\frac{2}{5} \times 60 = \underline{\quad}$ $\frac{6}{7} \times 56 = \underline{\quad}$

$\frac{5}{6} \times 42 = \underline{\quad}$ $\frac{5}{9} \times 54 = \underline{\quad}$ $\frac{3}{5} \times 35 = \underline{\quad}$ $\frac{5}{7} \times 63 = \underline{\quad}$

$\frac{5}{12} \times 36 = \underline{\quad}$ $\frac{5}{8} \times 56 = \underline{\quad}$ $\frac{4}{7} \times 35 = \underline{\quad}$ $\frac{2}{5} \times 25 = \underline{\quad}$

$\frac{7}{8} \times 96 = \underline{\quad}$ $\frac{4}{5} \times 90 = \underline{\quad}$ $\frac{3}{7} \times 28 = \underline{\quad}$ $\frac{3}{5} \times 40 = \underline{\quad}$

$\frac{3}{4} \times 32 = \underline{\quad}$ $\frac{5}{7} \times 84 = \underline{\quad}$ $\frac{4}{5} \times 50 = \underline{\quad}$ $\frac{4}{9} \times 63 = \underline{\quad}$

$\frac{5}{9} \times 36 = \underline{\quad}$ $\frac{5}{8} \times 56 = \underline{\quad}$ $\frac{4}{7} \times 35 = \underline{\quad}$ $\frac{2}{5} \times 25 = \underline{\quad}$

Maria and John made a pizza to share with friends. The girls ate $\frac{1}{4}$ of the pizza and the boys ate twice as much. What part did the boys eat? $\underline{\quad}$

James lives $\frac{5}{8}$ mile from school. How many miles does he ride his bicycle to and from school in a week? $\underline{\quad}$

FRACTIONS

MULTIPLYING WITH FRACTIONS.

Here are four fields. Each field is planted with fractions.
Help the farmer multiply the crops.

$\frac{3}{5}$ × 45 = _____

$\frac{3}{9}$ × 54 = _____

$\frac{4}{5}$ × 80 = _____

$\frac{2}{3}$ × 60 = _____

$\frac{2}{5}$ × 65 = _____

$\frac{3}{8}$ × 64 = _____

$\frac{2}{3}$ × 81 = _____

$\frac{3}{7}$ × 56 = _____

$\frac{2}{12}$ × 24 = _____

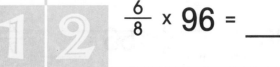

$\frac{6}{8}$ × 96 = _____

$\frac{3}{8}$ × 56 = _____

$\frac{2}{5}$ × 25 = _____

$\frac{9}{12}$ × 120 = _____

$\frac{5}{12}$ × 84 = _____

$\frac{5}{6}$ × 30 = _____

$\frac{2}{4}$ × 72 = _____

$\frac{10}{16}$ × 64 = _____

$\frac{1}{2}$ × 100 = _____

$\frac{2}{5}$ × 90 = _____

$\frac{2}{3}$ × 21 = _____

FRACTIONS

$\int$OLVING STORY PROBLEMS.

Complete the problems.

1. Phillip and his dad went fishing. They caught 20 fish, but they threw back all but $\frac{1}{4}$ of them. How many fish did they keep? _____

2. Loren, Mary, and Steven collected 15 boxes of toys to give to needy children. They sorted the toys and found that one-fifth of the boxes had broken toys. How many boxes of toys need to be fixed? _____

3. Mona and Junelle went to the Botanical Gardens. In the Butterfly Garden they counted 36 butterflies. $\frac{1}{9}$ of the butterflies were Monarchs. How many Monarch butterflies did Mona and Junelle see? _____

4. Joe and Frank built a clubhouse. They invited 28 of their friends to join. At the first meeting only $\frac{1}{7}$ of their friends came. How many came to the meeting? _____

5. The school band had ordered 42 new music stands. After 3 weeks, the school had received only $\frac{1}{7}$ of their order. How many music stands were still to arrive? _____

6. A "Keep The Park Green" Fund had $81.00 in its account. The members spent $\frac{1}{9}$ of the money cleaning up the park. How much was left in the account? _____

7. Chloe decided to sell pencils to make some extra money. She had 100 pencils and she sold them in packs of five. At the end of the day only $\frac{1}{10}$ of the packs were left. How many packs of pencils remained? _____

8. Jeniece and Maria bought a box of 27 cookies. When they opened the box, they saw that $\frac{1}{3}$ of them were broken up. How many cookies were broken? _____

19

COPYRIGHT © MILLIKEN PUBLISHING CO. MP4075

MULTIPLICATION OF FRACTIONS.

Write each product in its simplest form.

$\dfrac{7}{10} \times \dfrac{5}{6} = \underline{\quad}$ $\dfrac{5}{8} \times \dfrac{1}{4} = \underline{\quad}$ $\dfrac{2}{3} \times \dfrac{1}{9} = \underline{\quad}$

$\dfrac{1}{3} \times \dfrac{9}{10} = \underline{\quad}$ $\dfrac{1}{4} \times \dfrac{8}{11} = \underline{\quad}$ $\dfrac{8}{15} \times \dfrac{5}{6} = \underline{\quad}$ $\dfrac{3}{10} \times \dfrac{2}{3} = \underline{\quad}$

$\dfrac{6}{7} \times \dfrac{5}{6} = \underline{\quad}$ $\dfrac{3}{14} \times \dfrac{7}{12} = \underline{\quad}$ $\dfrac{7}{9} \times \dfrac{3}{7} = \underline{\quad}$ $\dfrac{4}{11} \times \dfrac{3}{16} = \underline{\quad}$

$\dfrac{2}{5} \times \dfrac{5}{9} = \underline{\quad}$ $\dfrac{7}{11} \times \dfrac{3}{14} = \underline{\quad}$ $\dfrac{1}{12} \times \dfrac{4}{5} = \underline{\quad}$ $\dfrac{3}{4} \times \dfrac{4}{7} = \underline{\quad}$

$\dfrac{1}{12} \times \dfrac{2}{5} = \underline{\quad}$ $\dfrac{4}{5} \times \dfrac{15}{16} = \underline{\quad}$ $\dfrac{9}{16} \times \dfrac{2}{3} = \underline{\quad}$ $\dfrac{3}{5} \times \dfrac{5}{12} = \underline{\quad}$

$\dfrac{3}{20} \times \dfrac{4}{9} = \underline{\quad}$ $\dfrac{8}{9} \times \dfrac{5}{12} = \underline{\quad}$ $\dfrac{3}{4} \times \dfrac{8}{15} = \underline{\quad}$ $\dfrac{2}{5} \times \dfrac{7}{12} = \underline{\quad}$

$\dfrac{9}{10} \times \dfrac{5}{18} = \underline{\quad}$ $\dfrac{3}{20} \times \dfrac{5}{9} = \underline{\quad}$ $\dfrac{2}{5} \times \dfrac{5}{8} = \underline{\quad}$ $\dfrac{5}{8} \times \dfrac{4}{15} = \underline{\quad}$

$\dfrac{2}{9} \times \dfrac{3}{5} \times \dfrac{5}{8} = \underline{\quad}$ $\dfrac{4}{5} \times \dfrac{5}{6} \times \dfrac{1}{2} = \underline{\quad}$ $\dfrac{3}{7} \times \dfrac{1}{3} \times \dfrac{7}{10} = \underline{\quad}$

$\dfrac{6}{7} \times \dfrac{3}{4} \times \dfrac{1}{3} = \underline{\quad}$ $\dfrac{8}{9} \times \dfrac{6}{11} \times \dfrac{11}{20} = \underline{\quad}$ $\dfrac{9}{10} \times \dfrac{4}{15} \times \dfrac{5}{6} = \underline{\quad}$

$\dfrac{5}{12} \times \dfrac{3}{4} \times \dfrac{2}{5} = \underline{\quad}$ $\dfrac{7}{20} \times \dfrac{3}{5} \times \dfrac{5}{7} = \underline{\quad}$ $\dfrac{1}{3} \times \dfrac{3}{8} \times \dfrac{2}{5} = \underline{\quad}$

1. Janet spends $\frac{2}{3}$ of her allowance on school lunches and $\frac{1}{6}$ on entertainment. What part of her allowance is left? _____

2. Shantel earned $2.50 babysitting. She spent $\frac{3}{5}$ of it on a book. How much did she pay for the book? _____

3. Bill received his $5.00 allowance on Friday. On Saturday he spent $2.00. What fraction of his allowance did he spend? _____

4. Tom spent $\frac{3}{4}$ of his $6.00 allowance on comic books. How much did he spend? _____

5. Anne, Gina, and Mary shared the cost of a $3.60 box of candy. If each girl paid the same amount, how much did each pay? _____

6. Ed spent $\frac{3}{5}$ of his $4.00 allowance. Carlos spent $\frac{3}{4}$ of his $3.00 allowance. Who spent the most? _____ How much more? _____

7. Nita saved $\frac{3}{8}$ of her $6.00 birthday gift. How much did she save? _____

8. Peggy borrowed $\frac{1}{6}$ of the cost of a $9.00 watch from her mother. How much did she borrow? _____

9. Alex earned $270.00 at his summer job. He spent $\frac{1}{3}$ on a bicycle, $\frac{2}{9}$ on a tape player, and $\frac{1}{6}$ on a fiberglass bow. How much did the bicycle cost? _____

10. Jasmine earns $140.00 each week. How much does she spend if $\frac{1}{4}$ of her money is deducted for taxes? _____; $\frac{1}{2}$ is spent on entertainment? _____; $\frac{1}{5}$ is spent on clothing? _____; and $\frac{1}{20}$ is put into savings? _____

11. Scott had $4.00. He spent $2.50 on a movie. What fraction of his money did he spend for the movie? _____

FRACTIONS

COOKING WITH FRACTIONS.

Write each product in its simplest form.

Make $\frac{1}{2}$ of this recipe.

COCONUT DROPS

$\frac{1}{2}$ x $\frac{3}{4}$ c. coco- = _____
nut

$\frac{1}{2}$ x **4** eggs = _____

$\frac{1}{2}$ x $\frac{3}{2}$ c. flour = _____

$\frac{1}{2}$ x $\frac{1}{3}$ tsp. salt = _____

$\frac{1}{2}$ x $\frac{3}{4}$ c. sugar = _____

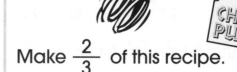

CHOCOLATE PUDDING

Make $\frac{2}{3}$ of this recipe.

$\frac{2}{3}$ x **8** c. milk = _____

$\frac{2}{3}$ x $\frac{7}{4}$ c. choc. = _____

$\frac{2}{3}$ x $\frac{6}{8}$ c. sugar = _____

$\frac{2}{3}$ x **3** eggs = _____

$\frac{2}{3}$ x **2** tsp. salt = _____

Make $\frac{3}{4}$ of this recipe.

POPCORN BALLS

$\frac{3}{4}$ x $\frac{7}{4}$ c. syrup = _____

$\frac{3}{4}$ x $\frac{2}{3}$ c. sugar = _____

$\frac{3}{4}$ x $\frac{1}{2}$ tsp. vanilla = _____

$\frac{3}{4}$ x $\frac{15}{2}$ c. popcorn = _____

$\frac{3}{4}$ x $\frac{5}{4}$ tsp. soda = _____

Make $\frac{5}{8}$ of this recipe.

CEREAL BARS

$\frac{5}{8}$ x $\frac{8}{3}$ c. cereal = _____

$\frac{5}{8}$ x $\frac{5}{4}$ c. marsh- = _____
mallows

$\frac{5}{8}$ x **2** sticks butter = _____

$\frac{5}{8}$ x $\frac{3}{4}$ tsp. vanilla = _____

$\frac{5}{8}$ x $\frac{2}{3}$ c. cocoa = _____

22

ANSWER KEY
FOR MILLIKEN'S
FRACTIONS WORKBOOK FOR

If you wish for the child to correct his or her own work, leave this answer key bound in the workbook for easy reference.

However, if you wish for someone else to correct the child's work, remove this answer key and keep it separate from the workbook. The key should separate easily, but if you feel it may tear, bend the staples out, slip the key out, and bend the staples back to keep the workbook intact.

Answers are set up in order of page number, by rows and columns, reading left to right.

PAGE 1:

$1\frac{1}{2}$ $\frac{6}{11}$ $\frac{4}{15}$ $1\frac{1}{3}$

$\frac{5}{7}$ $\frac{4}{7}$ $\frac{1}{4}$ $\frac{1}{10}$

$\frac{5}{8}$ $\frac{1}{3}$ $\frac{1}{9}$ $\frac{5}{8}$

$\frac{11}{12}$ $\frac{1}{16}$ $1\frac{7}{15}$ $\frac{1}{4}$

$1\frac{3}{20}$ $\frac{1}{2}$ $\frac{7}{8}$ $\frac{11}{12}$

PAGE 2:

$25\frac{1}{5}$

$21\frac{3}{4}$ $20\frac{3}{8}$ $19\frac{5}{8}$ $22\frac{7}{10}$

$24\frac{7}{12}$ $24\frac{11}{16}$ $127\frac{9}{16}$ $44\frac{13}{16}$

$116\frac{1}{4}$ $73\frac{7}{8}$ $129\frac{13}{16}$

$126\frac{5}{6}$

$144\frac{1}{8}$ $179\frac{1}{2}$ $106\frac{5}{12}$

$56\frac{5}{6}$

PAGE 3:

$72\frac{3}{8}$ $21\frac{5}{12}$

$31\frac{1}{2}$ $28\frac{1}{2}$ $20\frac{2}{3}$ $22\frac{5}{6}$

$93\frac{7}{8}$ $140\frac{3}{8}$ $27\frac{1}{12}$ $22\frac{1}{6}$

$27\frac{1}{9}$ $117\frac{5}{8}$ $39\frac{1}{3}$ $68\frac{2}{5}$

PAGE 3: (continued)

$86\frac{7}{8}$ $26\frac{23}{24}$ $68\frac{11}{12}$ $72\frac{9}{10}$

PAGE 4:

$12\frac{1}{7}$ $7\frac{4}{9}$ 12

$21\frac{1}{7}$ $22\frac{13}{14}$ $20\frac{3}{20}$ $8\frac{1}{8}$

29 $35\frac{19}{20}$ 21 $23\frac{1}{6}$

$24\frac{8}{11}$ $20\frac{7}{15}$ $13\frac{1}{21}$ 41

PAGE 5:

1. $30.00; $35.00

2. $\frac{5}{8}$

3. $28\frac{1}{2}$ minutes

4. $1\frac{1}{12}$ hours;

   $5\frac{7}{12}$ hours

5. $4\frac{2}{7}$ miles;

   $21\frac{3}{7}$ miles

6. $\frac{3}{4}$ $\frac{7}{9}$ $\frac{3}{5}$ $\frac{5}{6}$ $\frac{1}{3}$

7. $2\frac{3}{8}$ pounds

8. $11\frac{5}{8}$ pounds

PAGE 6:

$\frac{1}{3}$ $\frac{1}{2}$ $\frac{1}{3}$ $\frac{3}{4}$

PAGE 6: (continued)

$3\frac{3}{8}$ $12\frac{1}{8}$ $4\frac{1}{6}$

$12\frac{1}{4}$ $2\frac{1}{12}$ $5\frac{1}{2}$ $7\frac{3}{10}$

$6\frac{1}{8}$ $33\frac{1}{3}$ $2\frac{1}{2}$ $17\frac{3}{5}$

PAGE 7:

$14\frac{1}{4}$ $50\frac{2}{3}$ $3\frac{1}{8}$

$5\frac{3}{10}$ $38\frac{1}{3}$ $4\frac{19}{20}$

$1\frac{2}{3}$ $8\frac{3}{5}$ $8\frac{8}{15}$ $3\frac{5}{6}$

$10\frac{5}{8}$ $3\frac{2}{3}$ $1\frac{2}{3}$ $8\frac{1}{2}$

PAGE 8:

$38\frac{2}{7}$ $\frac{1}{3}$ $2\frac{3}{4}$

$3\frac{1}{2}$ $7\frac{3}{5}$ $4\frac{1}{2}$ $6\frac{1}{2}$

$1\frac{8}{9}$ $7\frac{1}{4}$ $8\frac{1}{2}$ $4\frac{1}{2}$

$3\frac{9}{10}$ $11\frac{3}{10}$ $2\frac{11}{14}$ $5\frac{5}{6}$

PAGE 9:

$25\frac{5}{8}$ $8\frac{2}{5}$ $25\frac{5}{6}$

$36\frac{5}{6}$ $5\frac{3}{8}$ $29\frac{3}{5}$ $6\frac{5}{12}$

$12\frac{1}{10}$ $18\frac{5}{7}$ $45\frac{5}{9}$ $6\frac{1}{2}$

$36\frac{2}{5}$ $34\frac{1}{7}$ $28\frac{1}{4}$ $8\frac{7}{8}$

$11\frac{3}{7}$ $9\frac{2}{3}$ $74\frac{4}{7}$ $27\frac{4}{5}$

$3\frac{3}{5}$ $3\frac{4}{7}$

$7\frac{1}{2}$ $1\frac{5}{7}$ $3\frac{4}{5}$ $5\frac{2}{5}$

$12\frac{2}{3}$ $6\frac{1}{2}$ $\frac{4}{5}$ $3\frac{1}{2}$

$12\frac{1}{4}$ $18\frac{2}{3}$ $\frac{3}{7}$ $4\frac{3}{7}$

$81\frac{1}{4}$ $7\frac{5}{6}$ $7\frac{4}{5}$ $6\frac{3}{7}$

$21\frac{5}{6}$ $7\frac{3}{7}$ $2\frac{3}{5}$ $4\frac{1}{2}$

$4\frac{3}{5}$ $2\frac{5}{8}$ $\frac{13}{16}$ $1\frac{13}{14}$

$82\frac{2}{3}$ $\frac{9}{10}$ $3\frac{2}{5}$ $1\frac{3}{4}$

$3\frac{4}{9}$ $17\frac{3}{4}$ $1\frac{3}{5}$ $2\frac{2}{7}$

$3\frac{4}{5}$ $9\frac{11}{12}$ $3\frac{4}{5}$ $9\frac{3}{4}$

$1\frac{7}{8}$ $11\frac{1}{2}$ $4\frac{1}{2}$ $21\frac{1}{5}$

$3\frac{11}{12}$ $6\frac{11}{15}$ $11\frac{9}{10}$ $3\frac{5}{8}$

$4\frac{4}{9}$ $4\frac{1}{3}$ $4\frac{7}{8}$ $2\frac{11}{18}$

$4\frac{13}{24}$ $7\frac{1}{12}$ $2\frac{1}{4}$ $3\frac{4}{5}$

$4\frac{7}{8}$ $3\frac{7}{10}$ $1\frac{7}{8}$ $3\frac{7}{12}$

$2\frac{9}{10}$ $2\frac{3}{4}$ $2\frac{1}{2}$ $5\frac{1}{2}$

$2\frac{4}{5}$ $1\frac{1}{2}$ $4\frac{5}{6}$ $1\frac{2}{3}$

$2\frac{7}{9}$ $1\frac{7}{8}$ $2\frac{13}{14}$ $\frac{3}{8}$

$1\frac{3}{5}$ $2\frac{5}{6}$ $5\frac{1}{3}$ $1\frac{9}{10}$

$3\frac{7}{10}$ $7\frac{7}{9}$ $\frac{1}{2}$ $1\frac{1}{4}$

$3\frac{11}{12}$ $4\frac{5}{8}$ $2\frac{2}{3}$ $4\frac{4}{5}$

$6\frac{1}{2}$ $3\frac{3}{5}$ $4\frac{5}{6}$ $2\frac{9}{14}$

$6\frac{1}{3}$

$\frac{2}{7}$ $\frac{7}{9}$

$\frac{3}{10}$ $\frac{4}{9}$

$\frac{7}{15}$ $\frac{1}{3}$

$\frac{3}{5}$ $\frac{1}{7}$

$\frac{2}{7}$ $\frac{2}{15}$ $\frac{3}{8}$ $\frac{2}{7}$

$4\frac{3}{5}$ $1\frac{1}{2}$ $7\frac{1}{3}$ $\frac{5}{7}$

$1\frac{3}{7}$ $2\frac{1}{5}$ $8\frac{1}{5}$ $2\frac{4}{5}$

$3\frac{2}{7}$ $14\frac{1}{4}$ $7\frac{1}{3}$ $7\frac{8}{9}$

1. $\frac{7}{8}$

2. $\frac{6}{12}$ or $\frac{1}{2}$

3. Barry; $\frac{1}{10}$

4. Yes; $\frac{6}{6}$ or 1

5. $\frac{5}{8}$ 6. $\frac{1}{4}$

7. Tyresha; $\frac{1}{10}$ mile

8. $1\frac{7}{16}$ cup   9. $\frac{1}{3}$

10. 4 pages

24	30	18	
36	45	28	54
12	18	24	48
35	30	21	45
15	35	20	10
84	72	12	24
24	60	40	28
20	35	20	10

$\frac{1}{2}$; $6\frac{1}{4}$ miles

1. 27, 18, 64, 40, 26
2. 24, 54, 24, 4, 72
3. 21, 10, 25, 36, 40
4. 90, 35, 50, 36, 14

1. 5 fish
2. 3 boxes

3. 4 butterflies
4. 4 friends
5. 36 music stands
6. $72.00
7. 2 packs
8. 9 cookies

$\frac{7}{12}$ $\frac{5}{32}$ $\frac{2}{27}$

$\frac{3}{10}$ $\frac{2}{11}$ $\frac{4}{9}$ $\frac{1}{5}$

$\frac{5}{7}$ $\frac{1}{8}$ $\frac{1}{3}$ $\frac{3}{44}$

$\frac{2}{9}$ $\frac{3}{22}$ $\frac{1}{15}$ $\frac{3}{7}$

$\frac{1}{30}$ $\frac{3}{4}$ $\frac{3}{8}$ $\frac{1}{4}$

$\frac{1}{15}$ $\frac{10}{27}$ $\frac{2}{5}$ $\frac{7}{30}$

$\frac{1}{4}$ $\frac{1}{12}$ $\frac{1}{4}$ $\frac{1}{6}$

$\frac{1}{12}$ $\frac{1}{3}$ $\frac{1}{10}$

$\frac{3}{14}$ $\frac{4}{15}$ $\frac{1}{5}$

$\frac{1}{8}$ $\frac{3}{20}$ $\frac{1}{20}$

1. $\frac{1}{6}$

2. $1.50

3. $\frac{2}{5}$

4. $4.50
5. $1.20
6. Ed; $.15 or 15¢
7. $2.25
8. $1.50
9. $90
10. $35; $70; $28; $7

11. $\frac{5}{8}$

Coconut Drops:
$\frac{3}{8}$ c.; 2 eggs; $\frac{3}{4}$ c.;

$\frac{1}{6}$ tsp.; $\frac{3}{8}$ c.

Chocolate Pudding:
$5\frac{1}{3}$ c.; $1\frac{1}{6}$ c.; $\frac{1}{2}$ c.;

2 eggs; $1\frac{1}{3}$ tsp.

Popcorn Balls:
$1\frac{5}{16}$ c.; $\frac{1}{2}$ c.; $\frac{3}{8}$ tsp.;

$5\frac{5}{8}$ c.; $\frac{15}{16}$ tsp.

Cereal Bars:
$1\frac{2}{3}$ c.; $\frac{25}{32}$ c.;

$1\frac{1}{4}$ stick; $\frac{15}{32}$ tsp.;

$\frac{5}{12}$ c.

PAGE 23:

1. 5 runners
2. 62 girls
3. $9\frac{3}{4}$ minutes
4. $51\frac{1}{2}$ minutes
5. 12 feet
6. 198 feet
7. 31 cans; $15.50; $4.50
8. 93 students

PAGE 24:

$\frac{1}{10}$ $\frac{1}{18}$

$\frac{1}{16}$ $\frac{2}{25}$ $\frac{4}{45}$

$\frac{1}{20}$ $\frac{1}{10}$ $\frac{5}{12}$

$\frac{1}{27}$ $\frac{2}{9}$ $\frac{1}{15}$

$\frac{11}{40}$ $\frac{1}{5}$ $\frac{1}{8}$

$\frac{5}{48}$ $\frac{3}{20}$ $\frac{1}{14}$

$\frac{1}{12}$ $\frac{1}{14}$ $\frac{1}{20}$

$\frac{1}{24}$ $\frac{4}{15}$ $\frac{1}{10}$

$\frac{7}{36}$ $\frac{4}{21}$ $\frac{3}{10}$

$\frac{2}{21}$ $\frac{2}{11}$ $\frac{7}{20}$

PAGE 25:

$1\frac{1}{7}$; $10\frac{1}{2}$

$\frac{4}{27}$; 7

PAGE 25: (cont)

$\frac{1}{8}$; $\frac{7}{9}$

$\frac{1}{5}$; $\frac{5}{32}$

21; $4\frac{1}{2}$

$1\frac{1}{4}$; 12

9; 8

16; $\frac{3}{8}$

4; 6

$\frac{2}{7}$; $\frac{5}{7}$

$\frac{2}{5}$; $\frac{1}{6}$

PAGE 26:

$2\frac{1}{2}$

$\frac{1}{2}$; $\frac{1}{3}$

$\frac{25}{32}$; $\frac{1}{3}$

$\frac{3}{4}$; 1

$1\frac{3}{7}$; $\frac{9}{10}$

$4\frac{2}{3}$; $\frac{2}{3}$

$\frac{3}{4}$; $\frac{1}{2}$

$\frac{9}{10}$; $1\frac{1}{3}$

2; $\frac{1}{2}$

3; $\frac{1}{2}$

$\frac{15}{16}$; $\frac{2}{3}$

$1\frac{1}{4}$

PAGE 27:

$1\frac{1}{4}$; $1\frac{3}{7}$

$2\frac{1}{2}$; $1\frac{1}{3}$

$\frac{4}{5}$; $\frac{14}{15}$

$\frac{7}{9}$; $\frac{3}{8}$

PAGE 27: (cont)

$\frac{8}{9}$; $1\frac{1}{3}$

$1\frac{1}{2}$; $\frac{9}{14}$

$\frac{3}{10}$; 2

$1\frac{2}{3}$; $1\frac{1}{5}$

$3\frac{1}{2}$; $\frac{2}{3}$

$1\frac{1}{15}$; $1\frac{1}{9}$

$1\frac{1}{4}$; 2

PAGE 28:

$\frac{1}{3}$; $\frac{5}{6}$

$\frac{25}{32}$; $\frac{1}{2}$; $1\frac{1}{9}$

$\frac{9}{10}$; $1\frac{5}{11}$; $1\frac{1}{5}$

$1\frac{1}{3}$; $\frac{5}{12}$; $1\frac{3}{7}$

1; $\frac{10}{27}$

$\frac{27}{64}$; $\frac{9}{50}$; $1\frac{1}{2}$

$\frac{12}{25}$; $\frac{1}{2}$; $1\frac{1}{48}$

$1\frac{1}{8}$; $1\frac{11}{16}$

$1\frac{5}{9}$; $1\frac{2}{3}$

PAGE 29:

$\frac{3}{2}$; $\frac{7}{8}$; $\frac{1}{2}$

$\frac{3}{4}$; $\frac{12}{5}$; $\frac{27}{14}$

$\frac{3}{4}$; $\frac{9}{8}$; $\frac{4}{5}$

$\frac{24}{5}$; 1; $\frac{7}{3}$

$\frac{1}{3}$; $\frac{1}{4}$; $\frac{3}{5}$

$\frac{5}{4}$; $\frac{9}{7}$; $\frac{3}{10}$

$\frac{5}{9}$; $\frac{5}{2}$; $\frac{3}{4}$

2; $\frac{3}{2}$; $\frac{4}{3}$

LA, KY, NY

PAGE 30:

$\frac{1}{6}$

55

$7\frac{1}{2}$; $\frac{1}{12}$

49; 28

$16\frac{1}{2}$; 2

$\frac{1}{8}$; 39

30; 8

$\frac{4}{9}$; $\frac{1}{6}$

54; 63

32; $3\frac{1}{2}$

$1\frac{3}{5}$; 48

PAGE 31:

$\frac{3}{8}$

8

$3\frac{3}{5}$; $\frac{2}{3}$

$\frac{3}{5}$; $\frac{1}{2}$

$\frac{2}{3}$; $\frac{2}{9}$

$\frac{1}{2}$; $\frac{1}{6}$

$3\frac{3}{4}$; $\frac{11}{14}$

$\frac{1}{2}$; 4

$1\frac{1}{2}$; 1

4; $\frac{2}{3}$

$\frac{1}{6}$; $\frac{3}{4}$

PAGE 32:

$8\frac{3}{4}$

22

$14\frac{2}{3}$

$14\frac{1}{3}$; $4\frac{4}{7}$

PAGE 32: (cont)

116; $20\frac{1}{2}$

56; 5

$7\frac{1}{2}$; $10\frac{1}{2}$

$11\frac{1}{5}$; $9\frac{1}{3}$

54; 15

21; $27\frac{1}{3}$

46; 108

$11\frac{1}{4}$; $12\frac{1}{4}$

PAGE 33:

$3\frac{3}{4}$

$1\frac{1}{6}$

$2\frac{1}{2}$; 4

4; $1\frac{7}{25}$

$2\frac{2}{3}$; $1\frac{1}{4}$

$1\frac{9}{10}$; $1\frac{1}{6}$

3; $\frac{25}{36}$

$\frac{2}{5}$; $2\frac{1}{2}$

$3\frac{3}{5}$; $4\frac{4}{5}$

$\frac{7}{8}$; $2\frac{1}{2}$

$\frac{1}{3}$; $1\frac{1}{2}$

2

PAGE 34:

$10\frac{1}{2}$

$3\frac{3}{4}$

$1\frac{6}{11}$; $4\frac{1}{3}$

$9\frac{1}{6}$; $4\frac{2}{7}$

$10\frac{1}{5}$; $19\frac{1}{4}$

$6\frac{3}{5}$; 2

$8\frac{1}{8}$; $1\frac{27}{28}$

$3\frac{3}{4}$; 3

4; $11\frac{7}{8}$

$7\frac{1}{2}$; $6\frac{2}{3}$

$5\frac{3}{4}$; $3\frac{4}{5}$

6

$4\frac{1}{8}$

$3\frac{3}{8}$

$2\frac{2}{3}$

$13\frac{1}{2}$

$11\frac{2}{3}$; $14\frac{2}{5}$

63; $6\frac{1}{2}$

$6\frac{3}{5}$; $25\frac{2}{3}$

14; $5\frac{1}{4}$

$3\frac{17}{21}$; $2\frac{2}{9}$

10; $10\frac{4}{5}$

9; 15

5; 2

Follow answers in target clockwise from upper right.

Top left target:

$3\frac{1}{8}$; $4\frac{1}{12}$; $12\frac{3}{5}$

$2\frac{3}{5}$; 3

Top right target:

6; $11\frac{1}{5}$; $5\frac{5}{12}$

$2\frac{1}{16}$; $12\frac{3}{8}$

Center target:

$6\frac{1}{2}$; $4\frac{7}{12}$; $1\frac{1}{3}$;

3; $8\frac{4}{5}$

Lower left target:

$1\frac{1}{3}$; $2\frac{4}{11}$; $3\frac{5}{7}$;

$2\frac{4}{5}$; $3\frac{1}{28}$

Lower right target:

$5\frac{13}{24}$; 5; $4\frac{1}{4}$;

$4\frac{22}{27}$; $3\frac{2}{11}$

$\frac{1}{8}$

4

10

$\frac{1}{2}$

$\frac{5}{16}$; $3\frac{1}{3}$

8; $1\frac{1}{12}$

$\frac{1}{9}$; $\frac{1}{12}$

$1\frac{1}{3}$; 5

$1\frac{5}{6}$; $\frac{3}{4}$

14; 8

$\frac{1}{4}$; $2\frac{2}{5}$

4; 12

$4\frac{1}{7}$; $1\frac{7}{8}$; $\frac{7}{18}$

$3\frac{1}{9}$; $2\frac{6}{13}$; $2\frac{1}{5}$

$\frac{7}{26}$; $1\frac{9}{22}$; $2\frac{4}{15}$

$1\frac{13}{35}$; $5\frac{5}{16}$; $\frac{18}{91}$

$\frac{2}{25}$; $\frac{1}{25}$; 18

$2\frac{14}{15}$; $9\frac{1}{2}$; 18

$\frac{7}{50}$; $\frac{5}{32}$; $\frac{5}{8}$

$1\frac{3}{4}$; $\frac{5}{12}$; $\frac{9}{10}$

$4\frac{2}{7}$; $3\frac{29}{33}$; 8

$\frac{2}{3}$

$\frac{6}{35}$

$\frac{2}{9}$

$\frac{25}{64}$; 6

$3\frac{1}{3}$; $8\frac{1}{6}$

$8\frac{2}{3}$; $\frac{4}{17}$

$\frac{3}{8}$; $13\frac{1}{2}$

$8\frac{2}{3}$; $\frac{7}{24}$

17; $1\frac{1}{3}$

$2\frac{1}{8}$; $5\frac{2}{3}$

$5\frac{1}{3}$; $\frac{15}{26}$

6; $\frac{1}{52}$

$1\frac{1}{8}$

$\frac{16}{17}$

$2\frac{1}{3}$

$1\frac{1}{2}$

$1\frac{2}{7}$; $2\frac{1}{2}$

$1\frac{3}{10}$; $\frac{9}{10}$

$3\frac{1}{9}$; $1\frac{1}{2}$

$\frac{5}{6}$; $\frac{1}{3}$

$1\frac{1}{20}$; $4\frac{1}{11}$

$1\frac{7}{9}$; $\frac{7}{9}$

$\frac{3}{4}$; 2

$1\frac{1}{5}$; $1\frac{2}{3}$

$1\frac{7}{8}$

$4\frac{4}{5}$

$\frac{5}{18}$

$\frac{3}{4}$

$1\frac{7}{9}$; $5\frac{1}{3}$

$\frac{2}{5}$; $1\frac{1}{9}$

$\frac{2}{5}$; 3

$1\frac{1}{5}$; $1\frac{2}{5}$

$\frac{14}{19}$; $6\frac{2}{3}$

$2\frac{1}{4}$; $3\frac{1}{5}$

$\frac{10}{21}$; $2\frac{8}{9}$

$\frac{7}{20}$; $\frac{21}{44}$

$\frac{3}{5}$

$2\frac{1}{2}$

$3\frac{1}{3}$

$\frac{2}{11}$; $\frac{1}{5}$

$2\frac{1}{16}$; $\frac{1}{14}$

$\frac{11}{12}$; 9

$5\frac{5}{12}$; 5

$4\frac{2}{5}$; $3\frac{1}{3}$

$4\frac{7}{8}$; 8

$\frac{2}{15}$; 2

4; $1\frac{1}{3}$

$2\frac{7}{10}$; $\frac{4}{5}$

$4\frac{4}{9}$

$1\frac{1}{4}$

$7\frac{7}{8}$

$1\frac{1}{99}$; $16\frac{1}{3}$

$\frac{1}{2}$; $4\frac{2}{5}$

7; $4\frac{5}{18}$

$\frac{7}{15}$; $1\frac{1}{2}$

$17\frac{1}{2}$; $\frac{18}{35}$

$2\frac{2}{3}$; $1\frac{11}{13}$

$\frac{1}{4}$; $7\frac{11}{12}$

6; 2

$\frac{15}{22}$

post office

$4\frac{7}{12}$; $6\frac{1}{2}$; $8\frac{2}{3}$

$6\frac{2}{5}$; $7\frac{4}{5}$; $4\frac{2}{5}$

$7\frac{1}{3}$; $14\frac{1}{5}$; $3\frac{1}{3}$

$9\frac{2}{3}$; $5\frac{1}{2}$; $7\frac{1}{2}$

$\frac{14}{15}$; $\frac{8}{9}$; $\frac{6}{11}$

2; $\frac{5}{11}$; $1\frac{5}{14}$

$\frac{9}{14}$; $2\frac{4}{7}$; $\frac{13}{22}$

$1\frac{1}{2}$; $5\frac{1}{10}$; $\frac{1}{2}$

$1\frac{5}{7}$; $3\frac{5}{8}$; $2\frac{11}{12}$

MEET THE JUNIOR SUPER STARS.

1. In the first race of the season, there were 15 runners. $\frac{1}{3}$ finished the race in less than 2 minutes. How many runners did this? _____

2. There were 124 participants in the track meet. Half of them were boys. How many were girls? _____

3. Maria finished her race in $3\frac{1}{4}$ minutes. Scott finished in $4\frac{1}{4}$ minutes. José finished in $2\frac{1}{4}$ minutes. How many total minutes to run all three races? _____

4. Tony was in every race. His total running times were $32\frac{1}{2}$ minutes. The team total was 84 minutes. How many minutes remain if Tony's are removed? _____

5. In Minh's three high jump tries he scored $4\frac{1}{2}$ feet, $4\frac{1}{8}$ feet, and $3\frac{3}{8}$ feet. How many total feet did he jump? _____

6. Manuel threw the discus three times. His scores were: 60 feet, 68 feet, and 70 feet. What was his total distance? _____

7. Georgia got refreshments for some of the 124 team members. She bought soda for $\frac{1}{4}$ of them. How many sodas did she buy? _____ If each soda cost 50¢, what was the total cost? _____
What was her change from a $20 bill? _____

8. John took a survey of the track team. He discovered $\frac{3}{4}$ of the team didn't like their new track uniforms. How many students didn't like their new suits? _____

23

MULTIPLICATION OF FRACTIONS.

Write each product in its simplest form.

$\dfrac{7}{8} \times \dfrac{3}{14} \times \dfrac{8}{15} =$ _____

$\dfrac{2}{5} \times \dfrac{5}{16} \times \dfrac{4}{9} =$ _____

$\dfrac{3}{8} \times \dfrac{4}{21} \times \dfrac{7}{8} =$ _____

$\dfrac{9}{20} \times \dfrac{4}{15} \times \dfrac{2}{3} =$ _____

$\dfrac{5}{6} \times \dfrac{8}{9} \times \dfrac{3}{25} =$ _____

$\dfrac{2}{5} \times \dfrac{3}{10} \times \dfrac{5}{12} =$ _____

$\dfrac{5}{18} \times \dfrac{2}{5} \times \dfrac{9}{10} =$ _____

$\dfrac{6}{7} \times \dfrac{5}{6} \times \dfrac{7}{12} =$ _____

$\dfrac{2}{15} \times \dfrac{5}{9} \times \dfrac{1}{2} =$ _____

$\dfrac{3}{4} \times \dfrac{1}{3} \times \dfrac{8}{9} =$ _____

$\dfrac{1}{4} \times \dfrac{8}{9} \times \dfrac{3}{10} =$ _____

$\dfrac{11}{12} \times \dfrac{4}{5} \times \dfrac{3}{8} =$ _____

$\dfrac{7}{20} \times \dfrac{4}{5} \times \dfrac{5}{7} =$ _____

$\dfrac{3}{5} \times \dfrac{5}{8} \times \dfrac{1}{3} =$ _____

$\dfrac{4}{11} \times \dfrac{5}{16} \times \dfrac{11}{12} =$ _____

$\dfrac{4}{15} \times \dfrac{9}{10} \times \dfrac{5}{8} =$ _____

$\dfrac{5}{12} \times \dfrac{2}{5} \times \dfrac{3}{7} =$ _____

$\dfrac{3}{8} \times \dfrac{5}{6} \times \dfrac{4}{15} =$ _____

$\dfrac{9}{16} \times \dfrac{4}{21} \times \dfrac{2}{3} =$ _____

$\dfrac{3}{8} \times \dfrac{1}{6} \times \dfrac{4}{5} =$ _____

$\dfrac{3}{20} \times \dfrac{4}{9} \times \dfrac{5}{8} =$ _____

$\dfrac{7}{9} \times \dfrac{4}{7} \times \dfrac{3}{5} =$ _____

$\dfrac{7}{10} \times \dfrac{3}{5} \times \dfrac{5}{21} =$ _____

$\dfrac{2}{5} \times \dfrac{7}{12} \times \dfrac{5}{6} =$ _____

$\dfrac{1}{4} \times \dfrac{8}{9} \times \dfrac{6}{7} =$ _____

$\dfrac{5}{12} \times \dfrac{4}{5} \times \dfrac{9}{10} =$ _____

$\dfrac{4}{9} \times \dfrac{1}{2} \times \dfrac{3}{7} =$ _____

$\dfrac{3}{4} \times \dfrac{1}{3} \times \dfrac{8}{11} =$ _____

$\dfrac{3}{5} \times \dfrac{2}{3} \times \dfrac{7}{8} =$ _____

24

DIVISION WITH FRACTIONS.

Write each quotient in its simplest form.

$2 \div \frac{7}{4} =$ _____

$6 \div \frac{4}{7} =$ _____

$\frac{4}{9} \div 3 =$ _____

$2 \div \frac{2}{7} =$ _____

$\frac{1}{2} \div 4 =$ _____

$\frac{7}{3} \div 3 =$ _____

$\frac{2}{5} \div 2 =$ _____

$\frac{5}{8} \div 4 =$ _____

$3 \div \frac{1}{7} =$ _____

$4 \div \frac{8}{9} =$ _____

$\frac{5}{2} \div 2 =$ _____

$6 \div \frac{1}{2} =$ _____

$8 \div \frac{8}{9} =$ _____

$6 \div \frac{3}{4} =$ _____

$8 \div \frac{1}{2} =$ _____

$\frac{3}{4} \div 2 =$ _____

$2 \div \frac{1}{2} =$ _____

$2 \div \frac{1}{3} =$ _____

$\frac{4}{7} \div 2 =$ _____

$\frac{10}{7} \div 2 =$ _____

$\frac{4}{5} \div 2 =$ _____

$\frac{5}{6} \div 5 =$ _____

FRACTIONS

DIVISION WITH FRACTIONS.

Write each quotient in its simplest form.

$\frac{5}{3} \div \frac{2}{3} = $ _____

$\frac{3}{8} \div \frac{3}{4} = $ _____

$\frac{1}{6} \div \frac{1}{2} = $ _____

$\frac{5}{8} \div \frac{4}{5} = $ _____

$\frac{1}{4} \div \frac{3}{4} = $ _____

$\frac{5}{8} \div \frac{5}{6} = $ _____

$\frac{1}{2} \div \frac{1}{2} = $ _____

$\frac{4}{7} \div \frac{2}{5} = $ _____

$\frac{3}{4} \div \frac{5}{6} = $ _____

$\frac{7}{2} \div \frac{3}{4} = $ _____

$\frac{1}{2} \div \frac{3}{4} = $ _____

$\frac{1}{8} \div \frac{1}{6} = $ _____

$\frac{1}{3} \div \frac{2}{3} = $ _____

$\frac{4}{5} \div \frac{8}{9} = $ _____

$\frac{1}{6} \div \frac{1}{8} = $ _____

$\frac{1}{4} \div \frac{1}{8} = $ _____

$\frac{1}{8} \div \frac{1}{4} = $ _____

$\frac{3}{4} \div \frac{1}{4} = $ _____

$\frac{1}{4} \div \frac{1}{2} = $ _____

$\frac{3}{4} \div \frac{4}{5} = $ _____

$\frac{1}{3} \div \frac{1}{2} = $ _____

$\frac{5}{6} \div \frac{2}{3} = $ _____

Write each quotient
in its simplest form.

$$\frac{5}{6} \div \frac{2}{3} = \frac{5}{\underset{2}{\cancel{6}}} \times \frac{\cancel{3}^{1}}{2} = \frac{5}{4} \text{ or } 1\frac{1}{4}$$

$\dfrac{5}{8} \div \dfrac{1}{2} = $ ____

$\dfrac{4}{7} \div \dfrac{2}{5} = $ ____

$\dfrac{5}{12} \div \dfrac{1}{6} = $ ____

$\dfrac{1}{6} \div \dfrac{1}{8} = $ ____

$\dfrac{1}{2} \div \dfrac{5}{8} = $ ____

$\dfrac{7}{9} \div \dfrac{5}{6} = $ ____

$\dfrac{7}{12} \div \dfrac{3}{4} = $ ____

$\dfrac{1}{4} \div \dfrac{2}{3} = $ ____

$\dfrac{4}{5} \div \dfrac{9}{10} = $ ____

$\dfrac{2}{5} \div \dfrac{3}{10} = $ ____

$\dfrac{9}{10} \div \dfrac{3}{5} = $ ____

$\dfrac{3}{8} \div \dfrac{7}{12} = $ ____

$\dfrac{1}{4} \div \dfrac{5}{6} = $ ____

$\dfrac{5}{7} \div \dfrac{5}{14} = $ ____

$\dfrac{5}{9} \div \dfrac{1}{3} = $ ____

$\dfrac{2}{3} \div \dfrac{5}{9} = $ ____

$\dfrac{7}{3} \div \dfrac{2}{3} = $ ____

$\dfrac{1}{3} \div \dfrac{1}{2} = $ ____

$\dfrac{4}{5} \div \dfrac{3}{4} = $ ____

$\dfrac{8}{9} \div \dfrac{4}{5} = $ ____

$\dfrac{5}{6} \div \dfrac{2}{3} = $ ____

$\dfrac{7}{6} \div \dfrac{7}{12} = $ ____

FIND THE QUOTIENTS.

$\frac{1}{4} \div \frac{3}{4} =$ _____

$\frac{1}{6} \div \frac{1}{5} =$ _____

$\frac{5}{8} \div \frac{4}{5} =$ _____

$\frac{3}{8} \div \frac{3}{4} =$ _____

$\frac{8}{9} \div \frac{4}{5} =$ _____

$\frac{3}{4} \div \frac{5}{6} =$ _____

$\frac{4}{11} \div \frac{2}{8} =$ _____

$\frac{2}{3} \div \frac{5}{9} =$ _____

$\frac{4}{5} \div \frac{9}{15} =$ _____

$\frac{1}{6} \div \frac{2}{5} =$ _____

$\frac{4}{7} \div \frac{2}{5} =$ _____

$\frac{1}{2} \div \frac{1}{2} =$ _____

$\frac{5}{30} \div \frac{45}{100} =$ _____

$\frac{3}{16} \div \frac{4}{9} =$ _____

$\frac{3}{20} \div \frac{5}{6} =$ _____

$\frac{9}{10} \div \frac{6}{10} =$ _____

$\frac{2}{5} \div \frac{5}{6} =$ _____

$\frac{3}{8} \div \frac{3}{4} =$ _____

$\frac{7}{12} \div \frac{4}{7} =$ _____

$\frac{3}{8} \div \frac{2}{6} =$ _____

$\frac{9}{16} \div \frac{1}{3} =$ _____

$\frac{2}{3} \div \frac{3}{7} =$ _____

$\frac{10}{12} \div \frac{1}{2} =$ _____

28

FRACTIONS

COPYRIGHT © MILLIKEN PUBLISHING CO. MP4075

DIVISION WITH FRACTIONS.

Write each quotient, then shade the answers.

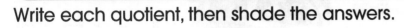

$\dfrac{3}{4} \div \dfrac{1}{2} =$ _____

$\dfrac{1}{3} \div \dfrac{4}{9} =$ _____

$\dfrac{1}{2} \div \dfrac{2}{3} =$ _____

$\dfrac{4}{5} \div \dfrac{1}{6} =$ _____

$\dfrac{1}{10} \div \dfrac{3}{10} =$ _____

$\dfrac{5}{6} \div \dfrac{2}{3} =$ _____

$\dfrac{5}{12} \div \dfrac{3}{4} =$ _____

$\dfrac{1}{4} \div \dfrac{1}{8} =$ _____

$\dfrac{7}{10} \div \dfrac{4}{5} =$ _____

$\dfrac{3}{5} \div \dfrac{1}{4} =$ _____

$\dfrac{5}{8} \div \dfrac{5}{9} =$ _____

$\dfrac{1}{2} \div \dfrac{1}{2} =$ _____

$\dfrac{1}{6} \div \dfrac{2}{3} =$ _____

$\dfrac{6}{7} \div \dfrac{2}{3} =$ _____

$\dfrac{5}{9} \div \dfrac{2}{9} =$ _____

$\dfrac{1}{2} \div \dfrac{1}{3} =$ _____

$\dfrac{1}{4} \div \dfrac{1}{2} =$ _____

$\dfrac{3}{7} \div \dfrac{2}{9} =$ _____

$\dfrac{2}{3} \div \dfrac{5}{6} =$ _____

$\dfrac{7}{9} \div \dfrac{1}{3} =$ _____

$\dfrac{3}{10} \div \dfrac{1}{2} =$ _____

$\dfrac{1}{5} \div \dfrac{2}{3} =$ _____

$\dfrac{3}{8} \div \dfrac{1}{2} =$ _____

$\dfrac{4}{9} \div \dfrac{1}{3} =$ _____

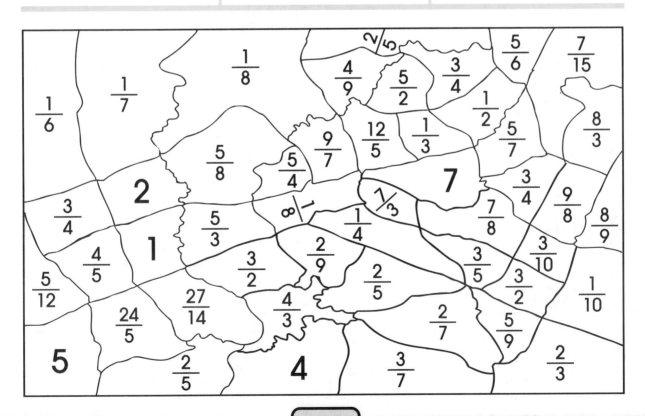

FRACTIONS

Write each product or quotient in its simplest form.

EXAMPLES:

$\frac{2}{3} \times 72 = \frac{2}{\underset{1}{\cancel{3}}} \times \frac{\overset{24}{\cancel{72}}}{1} = \mathbf{48}$

$4 \div \frac{8}{7} = \frac{\overset{1}{\cancel{4}}}{1} \times \frac{7}{\underset{2}{\cancel{8}}} = \frac{7}{2}$ or $\mathbf{3\frac{1}{2}}$

$\frac{5}{6} \div 5 =$ _____

$88 \times \frac{5}{8} =$ _____

$5 \div \frac{2}{3} =$ _____

$\frac{1}{3} \div 4 =$ _____

$84 \times \frac{7}{12} =$ _____

$\frac{7}{12} \times 48 =$ _____

$\frac{11}{16} \times 24 =$ _____

$3 \div \frac{3}{2} =$ _____

$\frac{7}{8} \div 7 =$ _____

$\frac{3}{5} \times 65 =$ _____

$\frac{3}{7} \times 70 =$ _____

$3 \div \frac{3}{8} =$ _____

$\frac{8}{9} \div 2 =$ _____

$\frac{1}{2} \div 3 =$ _____

$\frac{9}{16} \times 96 =$ _____

$\frac{7}{9} \times 81 =$ _____

$\frac{2}{5} \times 80 =$ _____

$4 \div \frac{8}{7} =$ _____

$2 \div \frac{5}{4} =$ _____

$\frac{2}{3} \times 72 =$ _____

Write each product or quotient in its simplest form.

EXAMPLES:

$$\frac{\overset{1}{\cancel{3}}}{\underset{2}{\cancel{14}}} \times \frac{\overset{1}{\cancel{7}}}{\underset{3}{\cancel{9}}} = \frac{1}{6}$$

$$\frac{1}{2} \div \frac{1}{8} = \frac{1}{\underset{1}{\cancel{2}}} \times \frac{\overset{4}{\cancel{8}}}{1} = 4$$

$\frac{2}{5} \times \frac{15}{16} = $ _____

$\frac{4}{5} \div \frac{1}{10} = $ _____

$\frac{6}{7} \div \frac{5}{21} = $ _____ $\frac{3}{5} \div \frac{9}{10} = $ _____

$\frac{9}{10} \times \frac{2}{3} = $ _____ $\frac{8}{9} \times \frac{9}{16} = $ _____

$\frac{4}{9} \div \frac{2}{3} = $ _____ $\frac{1}{12} \div \frac{3}{8} = $ _____

$\frac{5}{6} \times \frac{3}{5} = $ _____ $\frac{7}{9} \times \frac{3}{14} = $ _____

$\frac{5}{8} \div \frac{1}{6} = $ _____ $\frac{11}{12} \times \frac{6}{7} = $ _____

$\frac{6}{7} \times \frac{7}{12} = $ _____ $\frac{2}{3} \div \frac{1}{6} = $ _____

$\frac{1}{2} \div \frac{1}{3} = $ _____ $\frac{2}{5} \div \frac{2}{5} = $ _____

$\frac{1}{2} \div \frac{1}{8} = $ _____ $\frac{14}{15} \times \frac{5}{7} = $ _____

$\frac{3}{14} \times \frac{7}{9} = $ _____ $\frac{9}{10} \times \frac{5}{6} = $ _____

31

MULTIPLYING MIXED NUMBERS.

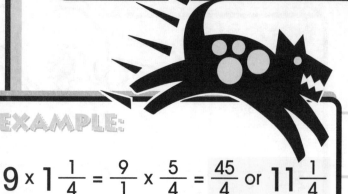

Write each product in its simplest form.

EXAMPLE:

$$9 \times 1\frac{1}{4} = \frac{9}{1} \times \frac{5}{4} = \frac{45}{4} \text{ or } 11\frac{1}{4}$$

$4\frac{3}{8} \times 2 = \underline{\hspace{2cm}}$

$1\frac{4}{7} \times 14 = \underline{\hspace{2cm}}$

$2\frac{4}{9} \times 6 = \underline{\hspace{2cm}}$

$2 \times 7\frac{1}{6} = \underline{\hspace{2cm}}$

$4 \times 1\frac{1}{7} = \underline{\hspace{2cm}}$

$8\frac{2}{7} \times 14 = \underline{\hspace{2cm}}$

$5 \times 4\frac{1}{10} = \underline{\hspace{2cm}}$

$5\frac{3}{5} \times 10 = \underline{\hspace{2cm}}$

$3 \times 1\frac{2}{3} = \underline{\hspace{2cm}}$

$7 \times 1\frac{1}{14} = \underline{\hspace{2cm}}$

$2\frac{5}{8} \times 4 = \underline{\hspace{2cm}}$

$4 \times 2\frac{4}{5} = \underline{\hspace{2cm}}$

$1\frac{5}{9} \times 6 = \underline{\hspace{2cm}}$

$10 \times 5\frac{2}{5} = \underline{\hspace{2cm}}$

$8 \times 1\frac{7}{8} = \underline{\hspace{2cm}}$

$6 \times 3\frac{1}{2} = \underline{\hspace{2cm}}$

$9\frac{1}{9} \times 3 = \underline{\hspace{2cm}}$

$3\frac{5}{6} \times 12 = \underline{\hspace{2cm}}$

$7\frac{1}{5} \times 15 = \underline{\hspace{2cm}}$

$9 \times 1\frac{1}{4} = \underline{\hspace{2cm}}$

$6\frac{1}{8} \times 2 = \underline{\hspace{2cm}}$

32

MULTIPLYING MIXED NUMBERS.

Write each product in its simplest form.

$\frac{5}{6} \times 4\frac{1}{2} =$ _____

EXAMPLE:

$2\frac{2}{3} \times \frac{3}{4} = \frac{\overset{2}{\cancel{8}}}{\underset{1}{\cancel{3}}} \times \frac{\overset{1}{\cancel{3}}}{\underset{1}{\cancel{4}}} = 2$

$\frac{2}{9} \times 5\frac{1}{4} =$ _____

$7\frac{1}{2} \times \frac{1}{3} =$ _____

$5\frac{1}{7} \times \frac{7}{9} =$ _____

$\frac{5}{8} \times 6\frac{2}{5} =$ _____

$\frac{4}{5} \times 1\frac{3}{5} =$ _____

$3\frac{1}{3} \times \frac{4}{5} =$ _____

$4\frac{1}{6} \times \frac{3}{10} =$ _____

$4\frac{3}{4} \times \frac{2}{5} =$ _____

$\frac{5}{9} \times 2\frac{1}{10} =$ _____

$\frac{6}{7} \times 3\frac{1}{2} =$ _____

$\frac{5}{8} \times 1\frac{1}{9} =$ _____

$3\frac{1}{5} \times \frac{1}{8} =$ _____

$\frac{3}{5} \times 4\frac{1}{6} =$ _____

$7\frac{1}{5} \times \frac{1}{2} =$ _____

$\frac{9}{10} \times 5\frac{1}{3} =$ _____

$\frac{1}{5} \times 4\frac{3}{8} =$ _____

$5\frac{5}{8} \times \frac{4}{9} =$ _____

$1\frac{1}{6} \times \frac{2}{7} =$ _____

$2\frac{1}{4} \times \frac{2}{3} =$ _____

$2\frac{2}{3} \times \frac{3}{4} =$ _____

33

FRACTIONS

MULTIPLYING MIXED NUMBERS.

Write each product in its simplest form.

EXAMPLE:

$$3\frac{8}{9} \times 1\frac{5}{7} = \frac{\overset{5}{\cancel{35}}}{\underset{3}{\cancel{9}}} \times \frac{\overset{4}{\cancel{12}}}{\underset{1}{\cancel{7}}} = \frac{20}{3} \text{ or } 6\frac{2}{3}$$

$6\frac{3}{4} \times 1\frac{5}{9} = $ _____

$1\frac{5}{12} \times 1\frac{1}{11} = $ _____

$2\frac{11}{12} \times 1\frac{2}{7} = $ _____

$4\frac{1}{6} \times 2\frac{1}{5} = $ _____

$3\frac{1}{4} \times 1\frac{1}{3} = $ _____

$3\frac{3}{5} \times 2\frac{5}{6} = $ _____

$2\frac{2}{7} \times 1\frac{7}{8} = $ _____

$2\frac{4}{9} \times 2\frac{7}{10} = $ _____

$4\frac{2}{3} \times 4\frac{1}{8} = $ _____

$3\frac{3}{4} \times 2\frac{1}{6} = $ _____

$1\frac{7}{11} \times 1\frac{2}{9} = $ _____

$2\frac{1}{12} \times 1\frac{4}{5} = $ _____

$1\frac{3}{7} \times 1\frac{3}{8} = $ _____

$2\frac{2}{3} \times 1\frac{1}{2} = $ _____

$2\frac{5}{8} \times 1\frac{1}{7} = $ _____

$2\frac{1}{10} \times 3\frac{4}{7} = $ _____

$6\frac{1}{4} \times 1\frac{9}{10} = $ _____

$5\frac{1}{9} \times 1\frac{1}{8} = $ _____

$3\frac{8}{9} \times 1\frac{5}{7} = $ _____

$1\frac{7}{12} \times 2\frac{2}{5} = $ _____

$4\frac{2}{5} \times 1\frac{4}{11} = $ _____

34

MULTIPLYING MIXED NUMBERS.

Write each product in its simplest form.

$1\frac{5}{16} \times 3\frac{1}{7} =$ _____

$2\frac{5}{11} \times 1\frac{3}{8} =$ _____

$2\frac{2}{9} \times 1\frac{1}{5} =$ _____

$2\frac{4}{7} \times 5\frac{1}{4} =$ _____

$1\frac{5}{9} \times 7\frac{1}{2} =$ _____ $2\frac{2}{15} \times 6\frac{3}{4} =$ _____

$8\frac{3}{4} \times 7\frac{1}{5} =$ _____ $3\frac{9}{10} \times 1\frac{2}{3} =$ _____

$4\frac{2}{5} \times 1\frac{1}{2} =$ _____ $2\frac{4}{5} \times 9\frac{1}{6} =$ _____

$5\frac{1}{3} \times 2\frac{5}{8} =$ _____ $4\frac{2}{3} \times 1\frac{1}{8} =$ _____

$2\frac{1}{7} \times 1\frac{7}{9} =$ _____ $2\frac{1}{12} \times 1\frac{1}{15} =$ _____

$2\frac{2}{9} \times 4\frac{1}{2} =$ _____ $2\frac{1}{10} \times 5\frac{1}{7} =$ _____

$5\frac{5}{8} \times 1\frac{3}{5} =$ _____ $3\frac{1}{3} \times 4\frac{1}{2} =$ _____

$2\frac{1}{17} \times 2\frac{3}{7} =$ _____ $1\frac{3}{11} \times 1\frac{4}{7} =$ _____

FRACTIONS

TARGETING FRACTIONS.

Multiply. Write the products as mixed numbers or whole numbers in the blank spaces.

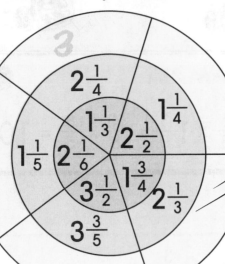

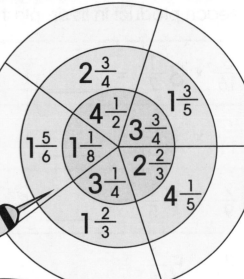

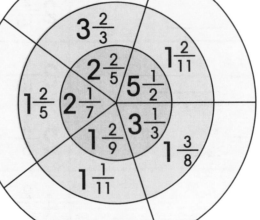

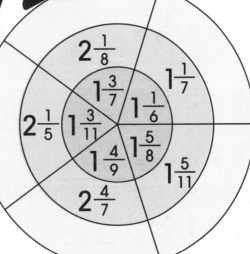

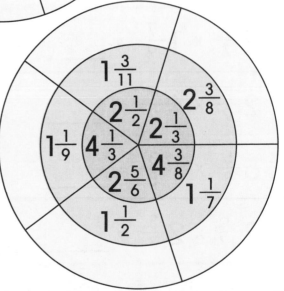

DIVIDING MIXED NUMBERS.

Write each quotient in its simplest form.

$2\frac{1}{8} \div 17 =$ _____

$14 \div 3\frac{1}{2} =$ _____

$25 \div 2\frac{1}{2} =$ _____

$2\frac{1}{2} \div 5 =$ _____

EXAMPLE:

$6 \div 1\frac{1}{2} = \frac{6}{1} \div \frac{3}{2} = \frac{\cancel{6}^{2}}{1} \times \frac{2}{\cancel{3}_{1}} = 4$

$1\frac{1}{4} \div 4 =$ _____ $5 \div 1\frac{1}{2} =$ _____

$15 \div 1\frac{7}{8} =$ _____ $3\frac{1}{4} \div 3 =$ _____

$1\frac{2}{3} \div 15 =$ _____ $2\frac{1}{2} \div 30 =$ _____

$2 \div 1\frac{1}{2} =$ _____ $8 \div 1\frac{3}{5} =$ _____

$3\frac{2}{3} \div 2 =$ _____ $3\frac{3}{4} \div 5 =$ _____

$30 \div 2\frac{1}{7} =$ _____ $20 \div 2\frac{1}{2} =$ _____

$1\frac{1}{4} \div 5 =$ _____ $4 \div 1\frac{2}{3} =$ _____

$6 \div 1\frac{1}{2} =$ _____ $42 \div 3\frac{1}{2} =$ _____

FRACTIONS

DIVIDING FRACTIONS.

Write each quotient in its simplest form.

$4\frac{5}{6} \div 1\frac{1}{6} =$ ___	$4\frac{5}{10} \div 2\frac{2}{5} =$ ___	$1\frac{2}{5} \div 3\frac{6}{10} =$ ___
$3\frac{1}{2} \div 1\frac{1}{8} =$ ___	$2\frac{2}{3} \div 1\frac{1}{12} =$ ___	$2\frac{4}{9} \div 1\frac{1}{9} =$ ___
$1\frac{1}{6} \div 4\frac{1}{3} =$ ___	$3\frac{1}{10} \div 2\frac{1}{5} =$ ___	$3\frac{2}{5} \div 1\frac{1}{2} =$ ___
$1\frac{1}{7} \div \frac{5}{6} =$ ___	$2\frac{1}{8} \div \frac{2}{5} =$ ___	$\frac{6}{7} \div 4\frac{1}{3} =$ ___
$\frac{1}{5} \div 2\frac{1}{2} =$ ___	$\frac{2}{8} \div 6\frac{1}{4} =$ ___	$3\frac{3}{5} \div \frac{2}{10} =$ ___
$2\frac{2}{10} \div \frac{3}{4} =$ ___	$3\frac{1}{6} \div \frac{1}{3} =$ ___	$4\frac{1}{2} \div \frac{1}{4} =$ ___
$1\frac{2}{5} \div 10 =$ ___	$1\frac{7}{8} \div 12 =$ ___	$1\frac{1}{4} \div 2 =$ ___
$9 \div 5\frac{1}{7} =$ ___	$1\frac{2}{3} \div 4 =$ ___	$4\frac{1}{2} \div 5 =$ ___
$6 \div 1\frac{2}{5} =$ ___	$16 \div 4\frac{1}{8} =$ ___	$20 \div 2\frac{1}{2} =$ ___

38

Write each quotient in its simplest form.

$\frac{5}{6} \div 1\frac{1}{4} =$ ____

EXAMPLE:

$5\frac{2}{5} \div \frac{9}{10} = \frac{27}{5} \div \frac{9}{10} =$

$\frac{\overset{3}{\cancel{27}}}{\underset{1}{\cancel{5}}} \times \frac{\overset{2}{\cancel{10}}}{\underset{1}{\cancel{9}}} = 6$

$\frac{4}{5} \div 4\frac{2}{3} =$ ____

$\frac{1}{3} \div 1\frac{1}{2} =$ ____

$\frac{5}{8} \div 1\frac{3}{5} =$ ____

$5\frac{1}{4} \div \frac{7}{8} =$ ____

$2\frac{2}{9} \div \frac{2}{3} =$ ____

$5\frac{5}{6} \div \frac{5}{7} =$ ____

$6\frac{1}{2} \div \frac{3}{4} =$ ____

$\frac{1}{2} \div 2\frac{1}{8} =$ ____

$\frac{7}{8} \div 2\frac{1}{3} =$ ____

$3\frac{3}{8} \div \frac{1}{4} =$ ____

$2\frac{3}{5} \div \frac{3}{10} =$ ____

$\frac{7}{9} \div 2\frac{2}{3} =$ ____

$7\frac{2}{7} \div \frac{3}{7} =$ ____

$1\frac{1}{9} \div \frac{5}{6} =$ ____

$1\frac{7}{10} \div \frac{4}{5} =$ ____

$2\frac{1}{8} \div \frac{3}{8} =$ ____

$3\frac{1}{5} \div \frac{3}{5} =$ ____

$\frac{3}{4} \div 1\frac{3}{10} =$ ____

$5\frac{2}{5} \div \frac{9}{10} =$ ____

$\frac{1}{10} \div 5\frac{1}{5} =$ ____

39

FRACTIONS

DIVIDING MIXED NUMBERS.

Write each quotient in its simplest form.

EXAMPLE:

$$3\frac{1}{2} \div 4\frac{1}{2} = \frac{7}{2} \div \frac{9}{2} =$$

$$\frac{7}{2} \times \frac{2}{9} = \boxed{\frac{7}{9}}$$

$2\frac{7}{10} \div 2\frac{2}{5} = \underline{\hspace{2cm}}$

$1\frac{3}{5} \div 1\frac{7}{10} = \underline{\hspace{2cm}}$

$3\frac{1}{3} \div 1\frac{3}{7} = \underline{\hspace{2cm}}$

$2\frac{1}{4} \div 1\frac{1}{2} = \underline{\hspace{2cm}}$

$2\frac{1}{7} \div 1\frac{2}{3} = \underline{\hspace{2cm}}$

$4\frac{3}{4} \div 1\frac{9}{10} = \underline{\hspace{2cm}}$

$1\frac{5}{8} \div 1\frac{1}{4} = \underline{\hspace{2cm}}$

$1\frac{2}{5} \div 1\frac{5}{9} = \underline{\hspace{2cm}}$

$5\frac{5}{6} \div 1\frac{7}{8} = \underline{\hspace{2cm}}$

$2\frac{1}{6} \div 1\frac{4}{9} = \underline{\hspace{2cm}}$

$3\frac{1}{8} \div 3\frac{3}{4} = \underline{\hspace{2cm}}$

$1\frac{1}{12} \div 3\frac{1}{4} = \underline{\hspace{2cm}}$

$1\frac{1}{6} \div 1\frac{1}{9} = \underline{\hspace{2cm}}$

$7\frac{1}{2} \div 1\frac{5}{6} = \underline{\hspace{2cm}}$

$4\frac{4}{9} \div 2\frac{1}{2} = \underline{\hspace{2cm}}$

$3\frac{1}{2} \div 4\frac{1}{2} = \underline{\hspace{2cm}}$

$2\frac{1}{10} \div 2\frac{4}{5} = \underline{\hspace{2cm}}$

$2\frac{2}{3} \div 1\frac{1}{3} = \underline{\hspace{2cm}}$

$4\frac{2}{5} \div 3\frac{2}{3} = \underline{\hspace{2cm}}$

$5\frac{1}{3} \div 3\frac{1}{5} = \underline{\hspace{2cm}}$

40

DIVIDING MIXED NUMBERS.

Write each quotient in its simplest form.

$2\frac{1}{12} \div 1\frac{1}{9} = $ _____

$5\frac{1}{7} \div 1\frac{1}{14} = $ _____

$1\frac{4}{9} \div 5\frac{1}{5} = $ _____

$3\frac{3}{8} \div 4\frac{1}{2} = $ _____

EXAMPLE:

$2\frac{2}{9} \div 4\frac{2}{3} = \frac{20}{9} \div \frac{14}{3} = $

$\dfrac{\overset{10}{\cancel{20}}}{\underset{3}{\cancel{9}}} \times \dfrac{\overset{1}{\cancel{3}}}{\underset{7}{\cancel{14}}} = \dfrac{10}{21}$

$3\frac{1}{9} \div 1\frac{3}{4} = $ _____ 　　　$6\frac{2}{5} \div 1\frac{1}{5} = $ _____

$1\frac{1}{6} \div 2\frac{11}{12} = $ _____ 　　　$6\frac{1}{9} \div 5\frac{1}{2} = $ _____

$1\frac{2}{3} \div 4\frac{1}{6} = $ _____ 　　　$6\frac{1}{2} \div 2\frac{1}{6} = $ _____

$3\frac{3}{10} \div 2\frac{3}{4} = $ _____ 　　　$3\frac{1}{5} \div 2\frac{2}{7} = $ _____

$1\frac{3}{4} \div 2\frac{3}{8} = $ _____ 　　　$7\frac{1}{3} \div 1\frac{1}{10} = $ _____

$5\frac{1}{4} \div 2\frac{1}{3} = $ _____ 　　　$4\frac{4}{5} \div 1\frac{1}{2} = $ _____

$2\frac{2}{9} \div 4\frac{2}{3} = $ _____ 　　　$3\frac{1}{4} \div 1\frac{1}{8} = $ _____

$1\frac{1}{10} \div 3\frac{1}{7} = $ _____ 　　　$2\frac{5}{8} \div 5\frac{1}{2} = $ _____

　　　FRACTIONS

MULTIPLICATION AND DIVISION.

Write each product or quotient in its simplest form.

EXAMPLES:

$$1\frac{3}{4} \div \frac{7}{8} = \frac{7}{4} \div \frac{7}{8} = \frac{\cancel{7}^1}{\cancel{4}_1} \times \frac{\cancel{8}^2}{\cancel{7}_1} = 2$$

$$4\frac{2}{3} \times \frac{2}{7} = \frac{\overset{2}{\cancel{14}}}{3} \times \frac{2}{\cancel{7}_1} = \frac{4}{3} \text{ or } 1\frac{1}{3}$$

$\frac{3}{8} \times 1\frac{3}{5} =$ _____

$3\frac{3}{4} \times \frac{2}{3} =$ _____

$\frac{2}{5} \times 8\frac{1}{3} =$ _____

$\frac{1}{2} \div 2\frac{3}{4} =$ _____

$\frac{1}{6} \times 1\frac{1}{5} =$ _____

$\frac{3}{8} \times 5\frac{1}{2} =$ _____

$\frac{1}{8} \div 1\frac{3}{4} =$ _____

$2\frac{3}{4} \times \frac{1}{3} =$ _____

$3\frac{3}{8} \div \frac{3}{8} =$ _____

$6\frac{1}{2} \times \frac{5}{6} =$ _____

$3\frac{3}{4} \div \frac{3}{4} =$ _____

$3\frac{2}{3} \div \frac{5}{6} =$ _____

$1\frac{1}{3} \div \frac{2}{5} =$ _____

$6\frac{1}{2} \times \frac{3}{4} =$ _____

$6\frac{2}{3} \div \frac{5}{6} =$ _____

$\frac{1}{5} \div 1\frac{1}{2} =$ _____

$1\frac{3}{4} \div \frac{7}{8} =$ _____

$5\frac{1}{3} \times \frac{3}{4} =$ _____

$4\frac{2}{3} \times \frac{2}{7} =$ _____

$2\frac{1}{4} \div \frac{5}{6} =$ _____

$1\frac{1}{5} \times \frac{2}{3} =$ _____

42

MULTIPLICATION AND DIVISION.

Write each product or quotient in its simplest form.

$1\frac{7}{9} \times 2\frac{1}{2} =$ _____

$5\frac{5}{6} \div 4\frac{2}{3} =$ _____

$3\frac{1}{2} \times 2\frac{1}{4} =$ _____

$2\frac{7}{9} \div 2\frac{3}{4} =$ _____ $4\frac{2}{3} \times 3\frac{1}{2} =$ _____

$2\frac{3}{4} \div 5\frac{1}{2} =$ _____ $2\frac{4}{5} \times 1\frac{4}{7} =$ _____

$4\frac{3}{8} \times 1\frac{3}{5} =$ _____ $1\frac{5}{6} \times 2\frac{1}{3} =$ _____

$2\frac{1}{10} \div 4\frac{1}{2} =$ _____ $3\frac{3}{5} \div 2\frac{2}{5} =$ _____

$8\frac{1}{6} \times 2\frac{1}{7} =$ _____ $1\frac{1}{14} \div 2\frac{1}{12} =$ _____

$2\frac{6}{7} \div 1\frac{1}{14} =$ _____ $1\frac{1}{11} \times 1\frac{9}{13} =$ _____

$3\frac{5}{8} \div 14\frac{1}{2} =$ _____ $4\frac{2}{9} \times 1\frac{7}{8} =$ _____

$3\frac{1}{3} \times 1\frac{4}{5} =$ _____ $7\frac{1}{2} \div 3\frac{3}{4} =$ _____

$2\frac{1}{2} \div 3\frac{2}{3} =$ _____

FRACTIONS

MULTIPLICATION AND DIVISION.

Write each product in its simplest form. Then decode the riddle.

WHAT TWO WORDS HAVE THE MOST LETTERS?

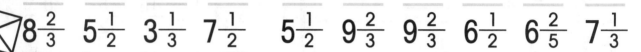

$8\frac{2}{3}$ $5\frac{1}{2}$ $3\frac{1}{3}$ $7\frac{1}{2}$ $5\frac{1}{2}$ $9\frac{2}{3}$ $9\frac{2}{3}$ $6\frac{1}{2}$ $6\frac{2}{5}$ $7\frac{1}{3}$

B $2\frac{3}{4} \times 1\frac{2}{3} = $ ___	**I** $1\frac{1}{3} \times 4\frac{7}{8} = $ ___	**P** $2\frac{3}{5} \times 3\frac{1}{3} = $ ___
C $3\frac{1}{5} \times 2 = $ ___	**K** $6\frac{1}{2} \times 1\frac{1}{5} = $ ___	**R** $5\frac{1}{2} \times \frac{4}{5} = $ ___
E $1\frac{3}{8} \times 5\frac{1}{3} = $ ___	**L** $2 \times 7\frac{1}{10} = $ ___	**S** $2\frac{1}{2} \times 1\frac{1}{3} = $ ___
F $4\frac{1}{7} \times 2\frac{1}{3} = $ ___	**O** $1\frac{3}{4} \times 3\frac{1}{7} = $ ___	**T** $3\frac{1}{2} \times 2\frac{1}{7} = $ ___

Write each quotient in its lowest terms.

$1\frac{3}{4} \div 1\frac{7}{8} = $ ___	$2\frac{1}{3} \div 2\frac{5}{8} = $ ___	$1\frac{1}{2} \div 2\frac{3}{4} = $ ___
$3\frac{2}{3} \div 1\frac{5}{6} = $ ___	$1\frac{1}{2} \div 3\frac{3}{10} = $ ___	$2\frac{3}{8} \div 1\frac{3}{4} = $ ___
$2\frac{1}{4} \div 3\frac{1}{2} = $ ___	$5\frac{2}{5} \div 2\frac{1}{10} = $ ___	$3\frac{1}{4} \div 5\frac{1}{2} = $ ___
$4\frac{1}{4} \div 2\frac{5}{6} = $ ___	$3\frac{2}{5} \div \frac{2}{3} = $ ___	$1\frac{2}{5} \div 2\frac{4}{5} = $ ___
$2\frac{3}{7} \div 1\frac{5}{12} = $ ___	$3\frac{5}{8} \div 1 = $ ___	$3\frac{3}{4} \div 1\frac{2}{7} = $ ___

44